BY LAURA LEE

Cosmic Connections: Exploring Humanity's Relationship with the Universe

Cosmic Connections: Exploring Humanity's Relationship with the Universe

By Laura Lee

Chapter 1: The Cosmic Perspective
- *Introducing the vastness of the universe and humanity's place within it.*
- *Discussing the historical and cultural perspectives on our relationship with the cosmos.*
- *Exploring the fundamental questions that arise from contemplating our existence in the universe.*

Chapter 2: The Origin Story
- *Delving into scientific theories of the universe's origin, from the Big Bang to cosmic evolution.*
- *Discussing how the elements that make up human bodies were forged in the depths of stars.*
- *Highlighting the interconnectedness between the cosmos and life on Earth.*

Chapter 3: Patterns in Nature
- *Exploring the patterns and structures found throughout the universe, from galaxies to DNA.*
- *Discussing the role of fractals and Fibonacci sequences in nature and human creations.*
- *Examining how humans have drawn inspiration from cosmic patterns in art, architecture, and design.*

Chapter 4: The Cosmic Web
- *Introducing the concept of the cosmic web, the vast network of galaxies interconnected by dark matter.*
- *Exploring the parallels between the cosmic web and networks in human society, such as the internet and social connections.*
- *Discussing the implications of the interconnectedness of the universe for our understanding of reality.*

Chapter 5: Consciousness and the Universe
- *Investigating the nature of consciousness and its relationship to the universe.*

Discussing the mysteries of quantum entanglement, superposition, and wave-particle duality.

Exploring the ways in which quantum physics challenges our notions of reality and expands our cosmic perspective.

Chapter 23: Cosmic Citizenship: Embracing Our Role in the Universe

Discussing the concept of cosmic citizenship and humanity's responsibilities as inhabitants of the cosmos.

Exploring how cosmic awareness can inspire environmental stewardship, social justice, and global cooperation.

Reflecting on the importance of cultivating a sense of cosmic citizenship in the face of planetary challenges and opportunities.

Chapter 24: The Symphony of the Cosmos: Exploring Cosmic Harmony

Exploring the concept of cosmic harmony and the interconnectedness of all things.

Discussing the role of chaos and order in the universe and in human society.

Examining how ancient and modern philosophies have sought to find meaning and purpose in the cosmos.

Chapter 25: The Cosmic Dance of Creation and Destruction

Exploring the dynamic interplay of creation and destruction in the cosmos.

Discussing the cosmic forces that shape the evolution of stars, galaxies, and planetary systems.

Reflecting on the parallels between cosmic processes and the cycles of change in human life and society.

Each chapter of "Cosmic Connections" delves deeper into the profound mysteries of the universe and explores the ways in which humanity is intimately connected to the cosmic tapestry that surrounds us. Through scientific inquiry, philosophical reflection, artistic expression, and spiritual exploration, we seek to uncover the deeper truths that unite us all in the grand cosmic dance of existence.

Prologue

In the quiet expanse of space, where stars flicker like distant beacons and galaxies swirl in cosmic dance, there exists a profound mystery that has captivated humanity since the dawn of time. It is a mystery that stretches beyond the boundaries of our understanding, beckoning us to venture ever deeper into the vast unknown.

In this prologue, we embark on a journey—a journey of exploration and discovery that will take us to the farthest reaches of the universe and back again. It is a journey that begins with a simple question: What is our place in the cosmos?

As we gaze upon the glittering tapestry of stars, we are filled with wonder and awe at the sheer magnitude of the universe. It is a universe of unimaginable scale and complexity—a universe that has been shaped by billions of years of cosmic evolution.

But amidst the vastness of space, there exists a remarkable truth: that we are not merely passive observers of the cosmos, but active participants in its grand symphony. For woven into the fabric of the universe are the same elements that make up our own bodies—the same atoms and molecules that have been forged in the fiery hearts of stars.

As we contemplate our cosmic origins, we are drawn into a deeper understanding of our interconnectedness with the cosmos. We are reminded that we are not separate from the universe, but rather, integral parts of its intricate web of existence.

In the pages that follow, we will explore the myriad ways in which humanity is connected to the universe—from the atoms in our bodies to the rhythms of cosmic time, from the poetry of the stars to the mysteries of dark matter and dark energy.

We will delve into the rich tapestry of human history and culture, uncovering the threads of cosmic wisdom that have been woven into the fabric of our collective consciousness. We will journey through the depths of space and time, guided by the light of curiosity and the fire of discovery.

And as we embark on this cosmic odyssey, we are filled with a sense of wonder and possibility–a recognition of the boundless potential that lies within each of us, waiting to be awakened by the infinite beauty of the universe.

So let us set forth, dear reader, with open hearts and open minds, into the vast expanse of the cosmos. For in the journey ahead, we will discover not only the mysteries of the universe, but also the profound truths that unite us all in the cosmic dance of existence.

Welcome to "Cosmic Connections: Exploring Humanity's Relationship with the Universe." The journey begins now.

The Cosmic Perspective

In the vast expanse of the universe, humanity is but a speck of dust, a tiny fragment of the cosmic tapestry that stretches across the cosmos. Yet, within this seemingly infinite realm, we have dared to ask some of the most profound questions that have ever crossed the human mind.

From the earliest civilizations to the present day, humans have looked to the heavens with wonder and awe, seeking to understand our place within the grandeur of the cosmos. In this chapter, we embark on a journey through time and space, exploring the myriad perspectives that humanity has held regarding our relationship with the universe.

Throughout history, cultures around the world have crafted intricate cosmologies and mythologies to explain the mysteries of the cosmos. From the ancient Greeks and Egyptians to the indigenous peoples of the

Americas and Africa, each culture has woven its own unique narrative of the heavens above, reflecting the values, beliefs, and aspirations of its people.

But beyond the realm of myth and legend lies a deeper truth–a truth that transcends the boundaries of culture and time. It is the truth that we are all connected, inextricably linked by the same forces that govern the universe. From the swirling galaxies to the tiniest particles of matter, we are bound together in a cosmic dance of existence, each one of us a thread in the tapestry of creation.

As we contemplate our place within the cosmos, we are confronted with fundamental questions that have puzzled philosophers, scientists, and theologians for centuries. What is the nature of existence? How did the universe come to be? And perhaps most profoundly, what is our purpose in the vastness of space and time?

These questions lie at the heart of the human experience, inviting us to explore the depths of our own consciousness and the mysteries of the universe. They challenge us to confront the limits of our understanding and to embrace the infinite possibilities that lie beyond.

As we gaze upon the stars, let us remember that we are not alone–that we are part of something greater than ourselves, bound together by the unbreakable threads of cosmic unity. And as we journey deeper into the cosmos, may we never lose sight of the wonder and awe that fills our hearts and minds, guiding us ever onward in our quest to unlock the secrets of the universe.

Welcome to the cosmic perspective–a journey of exploration, discovery, and wonder that knows no bounds. The adventure awaits.

In the grand narrative of cosmic history, the story of our universe unfolds like a majestic symphony, echoing across the vast expanse of space and time. From the fiery birth of the cosmos to the intricate dance of galaxies, the origin story of the universe is a tale of epic proportions.

At the heart of this story lies the Big Bang–a cataclysmic event that gave birth to the universe as we know it. In a fraction of a second, the cosmos exploded into existence, unleashing a torrent of energy and matter that would shape the course of cosmic evolution for billions of years to come.

As the universe expanded and cooled, subatomic particles coalesced into atoms, and atoms into stars and galaxies. Within the searing depths of these cosmic furnaces, the elements that make up our bodies–carbon, oxygen, nitrogen, and more–were forged in the fiery crucible of stellar fusion.

But the story does not end there. For billions of years, galaxies collided and stars died, scattering their elemental riches across the cosmos. And in the swirling clouds of gas and dust that permeated the cosmos, new stars and planets were born, giving rise to the cradle of life.

Here on Earth, amidst the vastness of space, life emerged–a wondrous tapestry of biodiversity, shaped by the forces of evolution and natural selection. From the smallest microbe to the largest mammal, life flourished and diversified, adapting to the ever-changing conditions of our planet.

But what does this cosmic saga mean for us, as inhabitants of planet Earth? It means that we are not separate from the universe, but rather, integral parts of its intricate web of existence. From the atoms in our bodies to the stars in the sky, we are bound together by the same cosmic forces that govern the universe.

As we contemplate our place in the cosmos, let us marvel at the interconnectedness of all things and the profound beauty of creation. For in the origin story of the universe, we find not only the mysteries of our cosmic heritage but also the promise of our shared destiny, bound together in the vast expanse of space and time.

Nature is a masterful artist, painting the canvas of the cosmos with intricate patterns and structures that mesmerize and inspire. From the swirling galaxies in the depths of space to the delicate petals of a flower, patterns abound in the natural world, revealing the beauty and complexity of the universe.

In this chapter, we embark on a journey through the patterns of nature, exploring their significance and their profound impact on human understanding and creativity.

At the grandest scales, galaxies dance in majestic spirals and elliptical arcs, their intricate shapes shaped by the forces of gravity and cosmic evolution. These cosmic patterns not only mesmerize astronomers and astrophysicists but also serve as a source of inspiration for artists, poets, and philosophers, who seek to capture the beauty and wonder of the universe in their creations.

But patterns are not confined to the vast reaches of space. They permeate every aspect of the natural world, from the branching veins of a leaf to the intricate structure of a snowflake. One of the most remarkable examples of patterns in nature is the Fibonacci sequence–a mathematical sequence that appears time and again in the spiral patterns of seashells, the arrangement of leaves on a stem, and the growth patterns of plants and flowers. The Fibonacci sequence is a testament to the underlying order and harmony of the universe, revealing the hidden mathematical principles that govern the natural world.

Another fascinating phenomenon found in nature is the presence of fractals–self-repeating geometric patterns that appear at different scales. From the branching patterns of trees to the intricate structure of river networks, fractals are ubiquitous in the natural world, reflecting the inherent complexity and beauty of nature's design.

But patterns are not only found in nature–they also permeate human creations, from the elegant curves of Gothic cathedrals to the intricate designs of Islamic tilework. Throughout history, humans have drawn inspiration from the patterns of the natural world, incorporating them into art, architecture, and design as a way of connecting with the cosmos and expressing the beauty and order of creation.

As we contemplate the patterns of nature, let us marvel at the intricate tapestry of the universe and the profound interconnectedness of all things. For in the patterns that surround us, we find not only beauty and order but also a reflection of the divine spark that animates the cosmos.

The Cosmic Web

In the vast expanse of the cosmos, a celestial network of unimaginable scale and complexity spans across the depths of space. This cosmic web, woven from the threads of dark matter and illuminated by the light of countless galaxies, is a marvel of cosmic architecture–a testament to the

interconnectedness of all things in the
universe.

At the heart of the cosmic web lies dark
matter—an invisible substance that
permeates the cosmos, exerting its
gravitational influence on the fabric of space
and time. Although dark matter remains
elusive and mysterious, its presence is
unmistakable, shaping the distribution of
galaxies and sculpting the cosmic landscape
on the largest scales.

But the cosmic web is more than just a
collection of galaxies—it is a vast network of
interconnected nodes, each one linked to its
neighbors by the invisible tendrils of dark
matter. Just as the neurons in our brains are
connected in a vast neural network, so too
are galaxies bound together in the cosmic
web, forming a complex and interconnected
system that spans across the universe.

In many ways, the cosmic web bears
striking similarities to networks in human
society, such as the internet and social
connections. Like the nodes of the cosmic
web, individuals and communities are linked

together by a web of relationships and interactions, forming a vast network of human connections that spans across the globe.

But what implications does the interconnectedness of the universe hold for our understanding of reality? As we peer into the depths of space and contemplate the cosmic web, we are confronted with profound questions about the nature of existence and our place in the cosmos. Are we merely isolated beings adrift in an indifferent universe, or are we part of something greater–a cosmic tapestry of interconnectedness and meaning?

The interconnectedness of the universe challenges our notions of reality, inviting us to expand our understanding and embrace the profound unity that binds us all together. For in the cosmic web, we find not only the beauty and diversity of the cosmos but also the deep interconnectedness of all things–a reminder that we are not separate from the universe, but integral parts of its intricate web of existence.

As we contemplate the cosmic web, let us marvel at the wonders of the universe and embrace the interconnectedness of all things. For in the cosmic tapestry that surrounds us, we find not only the mysteries of creation but also the promise of our shared destiny, bound together in the vast expanse of space and time.

Consciousness and the Universe

Consciousness, the enigmatic essence of our existence, is a profound mystery that has intrigued philosophers, scientists, and spiritual seekers for millennia. In this chapter, we embark on a journey to explore the nature of consciousness and its intricate relationship with the universe.

At its core, consciousness is the subjective experience of being aware of oneself and the world around us. From the simple awareness of our own thoughts and sensations to the complex depths of human emotion and perception, consciousness permeates every aspect of our existence, shaping our perceptions, beliefs, and actions.

But what is the nature of consciousness, and how does it relate to the universe? Philosophers have grappled with these questions for centuries, pondering the nature of mind and its relationship to the physical world. From the dualism of Descartes to the monism of Spinoza, different philosophical perspectives offer diverse insights into the nature of consciousness and its place in the cosmos.

In recent decades, scientists have also begun to explore the nature of consciousness from a scientific perspective, probing the neural correlates of consciousness and investigating the underlying mechanisms of subjective experience. From the intricate networks of neurons in the brain to the complex interplay of neurotransmitters and synaptic connections, scientists are beginning to unravel the mysteries of consciousness and its relationship to the physical world.

But beyond the confines of the individual mind lies a deeper truth–a truth that transcends the boundaries of the human experience. It is the truth that consciousness is not separate from the universe, but rather, an integral aspect of its intricate web of existence. From the cosmic dance of particles to the shimmering tapestry of galaxies, consciousness infuses every aspect of the universe, shaping its unfolding and evolution.

As we contemplate the nature of consciousness, let us embrace the profound interconnectedness of all things and the infinite possibilities that lie within the depths of our own awareness. For in the exploration of consciousness, we find not only the mysteries of the mind but also the profound wisdom of the universe, waiting to be discovered in the silent depths of our own being.

Time, the ever-present companion of our existence, weaves its intricate tapestry throughout the universe, shaping the rhythms of life and the unfolding of cosmic events. In this chapter, we embark on a journey to explore the concept of time—from the vast scales of the cosmos to the intimate depths of human perception.

At the cosmic level, time manifests as a fundamental aspect of the fabric of space-time—a dynamic framework that governs the evolution of the universe itself. From the fiery birth of stars to the relentless expansion of galaxies, time unfolds in a cosmic symphony of creation and transformation, shaping the destiny of the cosmos.

But time is not a fixed and immutable force —it is a fluid and dynamic phenomenon that varies with perspective and context. This truth lies at the heart of Einstein's theory of relativity, which revolutionized our understanding of time and space, revealing the profound interplay between matter, energy, and the curvature of space-time.

From the relativistic effects of gravity to the dilated perceptions of observers in motion, relativity challenges our intuitive notions of time and space, inviting us to explore the universe through the lens of curved geometry and warped dimensions.

But beyond the realm of physics lies a deeper truth–a truth that transcends the confines of scientific inquiry and touches the very essence of human experience. It is the truth that time is not merely a mechanical measure of duration, but a profound symbol of our existence–a fleeting moment in the cosmic dance of creation and dissolution.

Throughout history, cultures and religions around the world have conceptualized time in diverse and nuanced ways, reflecting the unique perspectives and insights of different societies. From the cyclical rhythms of nature in indigenous cultures to the linear progression of history in Western civilizations, time is understood and experienced in a multitude of ways, each one offering a window into the mysteries of the human spirit and the cosmos.

As we contemplate the cosmic perspectives on time, let us embrace the beauty and complexity of the universe and the profound mystery of our own existence. For in the exploration of time, we find not only the rhythms of the cosmos but also the timeless wisdom that transcends the bounds of space and eternity.

Cosmic Harmony

In the vast expanse of the cosmos, a symphony of harmony and chaos unfolds—a dance of order and disorder that shapes the fabric of existence itself. In this chapter, we embark on a journey to explore the concept of cosmic harmony and the profound interconnectedness of all things in the universe.

At its core, cosmic harmony is the recognition of the underlying unity that binds together the diverse elements of the cosmos. From the smallest subatomic particles to the largest galaxies, everything in the universe is interconnected, woven together in a seamless tapestry of existence.

But cosmic harmony is not merely a static state of equilibrium—it is a dynamic balance that emerges from the interplay of chaos and order. In the swirling depths of space, galaxies collide and stars are born and die, giving rise to the complex structures and patterns that define the cosmic landscape. From the turbulent storms of Jupiter to the delicate dance of planetary orbits, chaos and order coexist in a delicate balance, shaping the evolution of the universe.

In human society, the quest for cosmic harmony has been a central theme of philosophy, religion, and culture for millennia. From the ancient wisdom of the Greeks and Egyptians to the modern insights of scientists and thinkers, humanity has sought to find meaning and purpose in the cosmos, seeking to align our lives with the rhythms of the universe.

Ancient philosophies such as Taoism and Stoicism have emphasized the importance of living in harmony with nature and embracing the interconnectedness of all things. In the Eastern tradition, the concept of yin and yang symbolizes the balance between opposing forces, illustrating the dynamic interplay of light and dark, chaos and order, in the cosmic dance of existence.

In the Western tradition, philosophers such as Plato and Aristotle have explored the idea of the "harmony of the spheres"– the notion that the movements of the celestial bodies are governed by divine order and proportion, reflecting the harmony and beauty of the cosmos.

As we contemplate the concept of cosmic harmony, let us embrace the interconnectedness of all things and the profound wisdom of the universe. For in the recognition of our unity with the cosmos, we find not only meaning and purpose, but also the guiding light that illuminates our path in the journey of existence.

The Search for Extraterrestrial Life

Since time immemorial, humanity has gazed up at the stars, pondering the possibility of life beyond the confines of our own planet. In this chapter, we embark on a journey to explore the tantalizing prospect of extraterrestrial life and the scientific quest to uncover its existence.

Our fascination with the possibility of life beyond Earth stems from the profound question of whether we are alone in the universe. From ancient myths and legends to modern science fiction, the idea of extraterrestrial beings has captured the imagination of humanity, inspiring countless stories, myths, and dreams of cosmic exploration.

But beyond the realm of imagination lies the realm of scientific inquiry–a relentless quest to uncover the secrets of the cosmos and unlock the mysteries of life beyond Earth. From the exploration of Mars to the search for exoplanets in distant star systems, scientists are probing the depths of space in search of signs of extraterrestrial life.

In recent years, our understanding of the universe has been transformed by the discovery of thousands of exoplanets–worlds orbiting distant stars that bear tantalizing similarities to Earth. Some of these exoplanets lie within the "habitable zone" of their parent stars–the region where conditions may be right for the existence of liquid water and, potentially, life as we know it.

But the search for extraterrestrial life extends beyond the confines of our own solar system. Scientists are also exploring the possibility of life on other celestial bodies within our own solar system, such as the icy moons of Jupiter and Saturn, where liquid water may exist beneath the surface.

The discovery of extraterrestrial life would have profound implications for our understanding of humanity's place in the universe. It would not only expand our cosmic perspective but also challenge our notions of what it means to be alive and conscious in the vastness of space.

As we contemplate the search for extraterrestrial life, let us embrace the spirit of exploration and discovery that defines the human spirit. For in the quest to uncover the mysteries of the cosmos, we embark on a journey that transcends the boundaries of time and space, opening new horizons of knowledge and understanding that will shape the destiny of humanity for generations to come.

The Future of Cosmic Exploration

In the vast expanse of the cosmos, humanity stands at the threshold of a new era of exploration and discovery. In this chapter, we embark on a journey to explore the future of space exploration and humanity's potential to venture beyond the bounds of our own solar system.

The future of cosmic exploration holds immense promise and potential, driven by advances in technology, scientific inquiry, and human ambition. From the exploration of distant planets to the search for extraterrestrial life, humanity's quest to unlock the mysteries of the universe is poised to enter a new phase of discovery.

At the forefront of cosmic exploration is the exploration of Mars—a planet that has captured the imagination of scientists and explorers for centuries. With robotic missions already exploring the Martian surface and plans for crewed missions in the not-too-distant future, Mars represents a tantalizing destination for humanity's first steps beyond Earth.

But Mars is only the beginning. As technology continues to advance, humanity's reach will extend ever farther into the cosmos, with the potential to explore distant worlds, asteroid belts, and even neighboring star systems. From the development of advanced propulsion systems to the construction of space habitats and colonies, the future of cosmic exploration holds endless possibilities for humanity's expansion into the cosmos.

Yet, with great power comes great responsibility. As we venture into the unknown reaches of space, we must grapple with ethical and philosophical questions that will shape the course of cosmic exploration for generations to come. Questions of environmental sustainability, resource allocation, and the rights of indigenous peoples must be carefully considered as we embark on the journey to colonize other planets and encounter extraterrestrial civilizations.

But beyond the practical challenges lie deeper questions about the nature of humanity and our place in the universe. What does it mean to be human in a cosmos teeming with life and possibility? How will our encounters with other civilizations shape our understanding of ourselves and our place in the cosmos?

As we contemplate the future of cosmic exploration, let us embrace the spirit of exploration and discovery that defines the human spirit. For in the quest to unlock the mysteries of the universe, we embark on a journey that transcends the boundaries of time and space, shaping the destiny of humanity and the course of cosmic history for generations to come.

Finding Our Place in the Cosmos

As we reach the culmination of our cosmic journey, we find ourselves standing at the threshold of understanding, gazing out into the vast expanse of the universe with wonder and awe. In this final chapter, we reflect on the journey of exploration and discovery that has brought humanity to its current understanding of the cosmos, and we contemplate the profound significance of finding our place in the cosmic tapestry of existence.

Throughout history, humans have looked to the stars with a sense of wonder and curiosity, seeking to unlock the mysteries of the universe and uncover the secrets of our cosmic heritage. From the ancient astronomers who charted the movements of the heavens to the modern scientists who probe the depths of space with telescopes and spacecraft, the journey of cosmic exploration has been a testament to the boundless potential of human ingenuity and imagination.

But beyond the realm of scientific inquiry lies a deeper truth—a truth that transcends the boundaries of space and time and touches the very essence of our humanity. It is the truth that we are not separate from the universe, but rather, integral parts of its intricate web of existence. From the atoms in our bodies to the stars in the sky, we are connected to the cosmos in ways that we are only beginning to understand.

As we reflect on our place in the cosmos, let us embrace the profound interconnectedness of all things and cultivate a sense of reverence for the beauty and diversity of the universe. For in the recognition of our interconnectedness, we find not only meaning and purpose, but also a deep sense of belonging to something greater than ourselves.

In a world that often seems divided and fragmented, the cosmic perspective offers a powerful reminder of our shared humanity and our common destiny as inhabitants of planet Earth. It calls upon us to transcend the boundaries of race, religion, and nationality and to embrace the inherent unity that binds us together as members of the human family.

As we journey forward into the unknown depths of space and time, let us carry with us a sense of wonder and curiosity about the universe and our place within it. For in the exploration of the cosmos, we discover not only the mysteries of creation but also the infinite potential that lies within each of us, waiting to be awakened by the boundless beauty of the universe.

So let us gaze upon the stars with open hearts and open minds, and let us embrace the cosmic symphony that surrounds us, knowing that in the vastness of space, we are never truly alone. For in the journey of exploration and discovery, we find not only the secrets of the universe but also the profound wisdom that illuminates the path of our collective destiny.

As we navigate the vast expanse of the cosmos, let us remember that our journey of exploration extends not only outward, but inward as well. For in exploring the depths of our own consciousness and understanding, we unlock the secrets of the universe within ourselves.

In the interconnectedness of all things, we find solace and strength–a reminder that we are part of something greater than ourselves, bound together by the same forces that govern the cosmos. Let us cherish this interconnectedness and nurture a sense of stewardship for the precious planet we call home, recognizing that our actions ripple across the fabric of the universe.

As we stand on the threshold of the unknown, let us embrace the spirit of exploration and discovery that has defined humanity since the dawn of time. Let us seek out new horizons, new frontiers, and new possibilities, knowing that the journey of exploration is as much about the destination as it is about the path we take to get there.

And above all, let us never lose sight of the wonder and awe that fills our hearts and minds as we contemplate the vastness of the cosmos. For in the beauty of the stars and the mysteries of the universe, we find not only the answers to our questions but also the questions that have yet to be asked.

So let us continue to gaze upon the stars with wonder and curiosity, knowing that in the exploration of the cosmos, we discover not only the secrets of the universe but also the infinite potential that lies within each of us. For in the journey of finding our place in the cosmos, we find the true essence of what it means to be human.

Cosmic Inspiration in Art and Literature

Throughout the annals of human history, the universe has served as a boundless wellspring of inspiration for artists and writers, igniting the flames of creativity and imagination across cultures and civilizations.

In this chapter, we embark on a journey to explore the profound impact of cosmic themes on art and literature, delving into famous works that illuminate the wonders of the cosmos and reflect our timeless connection to the universe.

From the earliest cave paintings to the masterpieces of the Renaissance, artists have sought to capture the awe-inspiring beauty of the cosmos in their creations. The celestial bodies that adorn the night sky–the stars, planets, and galaxies–have served as symbols of divine majesty, stirring the hearts and minds of artists with their radiant beauty and timeless grace.

In literature, too, the universe has played a central role in shaping the narrative of human experience. From the epic poetry of Homer to the philosophical musings of Shakespeare, writers have drawn upon cosmic themes to explore the depths of human emotion and the mysteries of existence. Whether contemplating the vastness of space or the fleeting nature of time, literature has served as a mirror to the soul, reflecting the wonder and awe that fills our hearts as we gaze upon the night sky.

Some of the most iconic works of art and literature draw upon cosmic themes to convey timeless truths about the human condition. From Vincent van Gogh's swirling starry night to Jules Verne's visionary tales of space exploration, these creations speak to the universal longing for meaning and connection in a vast and mysterious universe.

But beyond the realm of aesthetics lies a deeper truth—a truth that transcends the boundaries of art and literature and touches the very essence of our humanity. It is the truth that we are not separate from the cosmos, but rather, integral parts of its intricate web of existence. In the creative expressions of artists and writers, we find echoes of our own longing for connection and understanding, reflecting the profound interconnectedness of all things in the universe.

As we reflect on the cosmic inspiration in art and literature, let us embrace the beauty and wonder of the universe that surrounds us. For in the exploration of creativity and imagination, we discover not only the secrets of the cosmos but also the infinite potential that lies within each of us, waiting to be unleashed by the boundless beauty of the universe.In the vast expanse of artistic expression, the cosmos serves as a limitless canvas upon which humanity projects its deepest hopes, fears, and aspirations. From the haunting melodies of symphonies inspired by the stars to the intricate brushstrokes of paintings that capture the majesty of the Milky Way, the universe has inspired countless works of art that evoke wonder and awe in those who behold them.

Literature, too, has served as a vessel for exploring the mysteries of the cosmos and the human experience within it. From the epic sagas of ancient civilizations to the speculative visions of science fiction writers, literature offers a window into the depths of the human soul and the boundless expanse of the universe.

Reflecting on the ways in which human creativity reflects our connection to the cosmos, we find that art and literature serve as bridges that span the vast chasm between the finite and the infinite, the mundane and the sublime. In the act of creation, artists and writers tap into a wellspring of inspiration that flows from the depths of the universe itself, channeling its energy and vitality into works that transcend the limitations of time and space.

But perhaps the most profound aspect of cosmic inspiration in art and literature lies in its ability to evoke a sense of wonder and curiosity about the universe and our place within it. As we immerse ourselves in the timeless beauty of a painting or lose ourselves in the pages of a novel, we are reminded of the vastness of existence and the endless possibilities that lie beyond the confines of our everyday lives.

In the end, cosmic inspiration in art and literature serves as a reminder of our shared humanity and our collective journey through the cosmos. It invites us to embrace the beauty and complexity of the universe and to celebrate the creative spirit that binds us together as members of the human family.

As we continue to explore the wonders of the cosmos, let us draw inspiration from the rich tapestry of artistic expression that surrounds us. For in the beauty of art and literature, we find not only the reflections of our own humanity but also the echoes of the stars themselves, shining brightly in the vast expanse of space and time.

The Mysteries of Dark Matter and Dark Energy

In the depths of the cosmos lie two enigmatic forces that hold the keys to understanding the very fabric of the universe–dark matter and dark energy. In this chapter, we embark on a journey to investigate these mysterious phenomena, exploring their profound influence on the structure and evolution of the cosmos and delving into the ongoing scientific efforts to unravel their secrets.

Dark matter, the invisible substance that pervades the universe, remains one of the most elusive mysteries of modern astrophysics. Unlike ordinary matter, which emits light and can be detected through traditional means, dark matter reveals itself only through its gravitational effects on visible matter and light. Yet despite its pervasive presence, the true nature of dark matter remains shrouded in mystery, tantalizing scientists with its hidden secrets.

But what role does dark matter play in shaping the structure and evolution of the universe? From the formation of galaxies to the dynamics of cosmic expansion, dark matter exerts its gravitational influence on the cosmic landscape, sculpting the vast structures that define the cosmos. Without dark matter, galaxies would not have enough mass to hold themselves together, and the universe as we know it would look vastly different.

Yet dark matter is only part of the cosmic puzzle. In recent decades, astronomers have discovered another mysterious force at work in the cosmos–dark energy. Unlike dark matter, which acts as a gravitational glue that binds galaxies together, dark energy behaves as a cosmic repulsive force, driving the accelerated expansion of the universe.

The discovery of dark energy has revolutionized our understanding of cosmic evolution, challenging traditional models of the universe and posing profound questions about its ultimate fate. What is the nature of dark energy, and how does it interact with the fabric of space and time? Is dark energy a fundamental property of the universe, or is it indicative of new physics waiting to be discovered?

These questions lie at the forefront of modern cosmology, driving scientific efforts to unravel the mysteries of dark matter and dark energy. From particle physics experiments deep underground to precision cosmological observations with cutting-edge telescopes, scientists are pushing the boundaries of human knowledge in search of answers.

As we delve deeper into the mysteries of dark matter and dark energy, we are reminded of the boundless complexity and wonder of the universe. For in the exploration of the unknown, we find not only the answers to our questions but also new questions waiting to be asked, leading us ever closer to unlocking the secrets of the cosmos and our place within it. As scientists continue their quest to understand dark matter and dark energy, new theories and experimental approaches are constantly being developed. Particle physicists are exploring the possibility of detecting dark matter particles directly or indirectly, hoping to shed light on their elusive nature. Cosmologists are using sophisticated telescopes and observational techniques to map the distribution of dark matter and study its effects on the large-scale structure of the universe.

Similarly, the study of dark energy is advancing rapidly, with astronomers using powerful observatories to observe distant supernovae and map the cosmic microwave background radiation in order to better understand the universe's expansion history. These efforts are essential for gaining insights into the fundamental properties of dark energy and its implications for the fate of the cosmos.

Yet, despite decades of research, many questions about dark matter and dark energy remain unanswered. What is the fundamental nature of these mysterious substances? How do they interact with ordinary matter and energy? And what role do they play in the cosmic drama of creation and destruction?

As we confront these profound mysteries, we are reminded of the inherent limitations of human knowledge and the boundless depths of the universe. Dark matter and dark energy stand as reminders of the vast unknown that lies beyond the reach of our understanding, challenging us to push the boundaries of scientific inquiry and expand the horizons of human knowledge.

In the face of such daunting challenges, we are called upon to embrace the spirit of curiosity and exploration that defines the human spirit. For in the pursuit of knowledge, we find not only the answers to our questions but also the joy of discovery and the thrill of exploration. And as we journey ever deeper into the mysteries of the cosmos, we are bound together by a shared sense of wonder and awe at the vastness and complexity of the universe that surrounds us.

Cosmic Evolution and the Fate of the Universe

In the vast expanse of cosmic time, the universe has undergone a remarkable journey of evolution, transforming from a primordial sea of energy into the rich tapestry of galaxies, stars, and planets that we observe today. In this chapter, we trace the epic saga of cosmic evolution from its earliest moments to the present day, and explore the profound implications of the universe's ultimate fate for humanity and the cosmos.

The story of cosmic evolution begins nearly 13.8 billion years ago with the Big Bang–a cataclysmic event that marked the birth of the universe and the beginning of space and time. In the fiery crucible of the early universe, matter and energy danced in a primordial soup of particles, gradually coalescing into the first atoms and molecules that would form the building blocks of galaxies and stars.

Over billions of years, gravity sculpted the cosmic landscape, drawing matter into vast filaments and clusters that spanned the expanse of space. Stars ignited and died in spectacular explosions, seeding the cosmos with the elements necessary for the formation of planets and the emergence of life.

But as the universe expands and evolves, so too do the forces that shape its destiny. Theories of the universe's ultimate fate abound, ranging from scenarios of continued expansion to the possibility of contraction or collapse. Some cosmologists predict a future in which the universe continues to expand indefinitely, with galaxies drifting farther and farther apart until they fade from view.

Others speculate that the universe may undergo a process of contraction, collapsing under the force of gravity into a fiery inferno known as the Big Crunch. Still, others propose the existence of exotic phenomena such as dark energy, which could drive an accelerated expansion of the cosmos and ultimately lead to a "Big Freeze" as the universe grows cold and dark.

The implications of cosmic evolution for humanity's long-term future are profound and far-reaching. As we contemplate the fate of the universe, we are confronted with existential questions about our place in the cosmos and the ultimate destiny of our species. Will humanity one day venture beyond the bounds of our solar system and colonize distant worlds? Or will we remain forever bound to the cradle of Earth, gazing out into the endless expanse of space with wonder and awe?

In the face of such uncertainty, one thing remains clear: the story of cosmic evolution is still being written, and humanity has a vital role to play in shaping its outcome. Whether exploring the mysteries of the cosmos or working to safeguard the future of our planet, we are called upon to embrace the spirit of discovery and exploration that defines the human experience.

As we journey forward into the unknown depths of space and time, let us carry with us a sense of wonder and curiosity about the universe and our place within it. For in the exploration of cosmic evolution, we find not only the secrets of the cosmos but also the boundless potential of the human spirit, reaching out to touch the stars and shape the destiny of the universe itself.

In the grand tapestry of cosmic evolution, humanity emerges as a pivotal protagonist, a species endowed with the capacity for understanding and exploration. In this chapter, we delve into humanity's cosmic journey, tracing our humble origins on planet Earth to our bold ventures into the depths of space and time.

Our story begins millions of years ago, when our distant ancestors first walked the savannas of Africa, gazing up at the same stars that still shine in the night sky today. Over countless generations, humanity's curiosity and ingenuity propelled us forward, driving us to explore new territories, discover new truths, and push the boundaries of knowledge and understanding.

From the ancient astronomers who charted the movements of the heavens to the modern scientists who probe the mysteries of the cosmos with powerful telescopes and spacecraft, humanity's quest for knowledge has taken us to the farthest reaches of the universe and the deepest depths of the atom.

But our journey is not just one of scientific inquiry—it is also a journey of self-discovery, as we grapple with the profound questions of existence and purpose that lie at the heart of the human experience. What does it mean to be human in a universe teeming with life and possibility? What is our place in the cosmic order, and what role do we play in shaping the destiny of the cosmos?

As we explore these questions, we are reminded of the interconnectedness of all things and the profound responsibility that comes with being stewards of the Earth and guardians of the universe. In the face of environmental challenges and existential threats, humanity must unite in common purpose to safeguard the future of our planet and ensure the survival of future generations.

Yet, even as we confront the challenges that lie ahead, we are filled with a sense of wonder and possibility at the vastness and beauty of the cosmos. In the twinkling of distant stars and the swirling galaxies that adorn the night sky, we find inspiration and hope for a future filled with exploration, discovery, and adventure.

As we continue our cosmic journey, let us embrace the spirit of exploration and discovery that defines the human experience. Let us reach for the stars with open hearts and open minds, knowing that in the exploration of the cosmos, we discover not only the secrets of the universe but also the infinite potential that lies within each of us, waiting to be unleashed by the boundless beauty of the cosmos.

Cosmic Connections in Indigenous Cultures

In the rich tapestry of human experience, indigenous cultures around the world offer profound insights into our interconnectedness with the cosmos. In this chapter, we embark on a journey to explore the cosmologies of indigenous peoples, discussing the ways in which their traditional knowledge and beliefs reflect humanity's deep connection to the natural world. We also examine the importance of preserving indigenous perspectives on the cosmos in the face of modernization and globalization.

Indigenous cultures have inhabited every corner of the Earth for thousands of years, developing unique cosmologies that reflect their intimate relationship with the land, the sky, and the creatures that inhabit them. From the Aboriginal peoples of Australia to the Inuit of the Arctic and the Native Americans of North America, indigenous cosmologies are steeped in a deep reverence for the natural world and the cycles of life and death that govern it.

At the heart of indigenous cosmologies lies a profound understanding of the interconnectedness of all things. For indigenous peoples, the Earth is not merely a resource to be exploited but a living, breathing entity with its own consciousness and spirit. Every tree, every rock, every creature is imbued with a sacred essence, and every action has consequences that ripple through the web of life.

In indigenous cosmologies, the sky is often seen as a mirror of the Earth, reflecting the patterns and rhythms of the natural world. The movements of the stars, the phases of the moon, and the cycles of the seasons are all seen as interconnected aspects of a greater cosmic dance–a dance that we are all a part of, whether we realize it or not.

Traditional knowledge and beliefs passed down through generations serve as a guiding light for indigenous peoples, offering wisdom and insight into the mysteries of the cosmos and the human spirit. Yet, in the face of modernization and globalization, many indigenous cultures are facing unprecedented challenges to their way of life, as traditional lands are encroached upon, languages are lost, and sacred traditions are threatened.

As we reflect on the importance of preserving indigenous perspectives on the cosmos, we are reminded of the vital role that diversity and cultural heritage play in shaping the fabric of human civilization. The wisdom of indigenous peoples offers valuable lessons for humanity as we navigate the challenges of the 21st century, reminding us of the interconnectedness of all life and the importance of living in harmony with the natural world.

In the spirit of reconciliation and mutual respect, let us embrace the teachings of indigenous cultures and work together to preserve their rich cultural heritage for generations to come. For in the preservation of indigenous perspectives on the cosmos, we find not only the wisdom of the ages but also a profound reminder of our shared humanity and our collective responsibility to protect the Earth and all its inhabitants.

As we continue to explore the cosmologies of indigenous cultures, we uncover a wealth of knowledge and wisdom that offers profound insights into our place in the cosmos. Indigenous peoples have long recognized that humanity is intricately connected to the natural world, and their cosmologies reflect this fundamental truth in myriad ways.

Central to many indigenous cosmologies is the concept of reciprocity—the idea that humans are not separate from nature, but rather, part of a vast network of relationships that extends to all living beings. In indigenous traditions, humans are regarded as caretakers of the Earth, entrusted with the responsibility of preserving the balance and harmony of the natural world.

For indigenous peoples, the cosmos is not simply a distant realm beyond reach; it is an integral part of everyday life, woven into the fabric of community, culture, and spirituality. The sun, moon, stars, and planets are not just celestial bodies; they are living entities with their own stories, teachings, and significance.

Through rituals, ceremonies, and oral traditions, indigenous cultures pass down their knowledge of the cosmos from generation to generation, ensuring that their connection to the natural world remains strong. These teachings offer valuable lessons in humility, respect, and gratitude, reminding us of the importance of living in harmony with the Earth and all its inhabitants.

However, the rapid pace of modernization and globalization has placed immense pressure on indigenous cultures and their traditional ways of life. Indigenous peoples face threats to their lands, resources, and cultural heritage, as well as challenges to their spiritual beliefs and practices. As a result, many indigenous communities are fighting to protect their rights, preserve their cultures, and safeguard their ancestral knowledge for future generations.

In the face of these challenges, it is imperative that we listen to and learn from indigenous voices, honoring their wisdom and resilience in the face of adversity. By recognizing the value of indigenous perspectives on the cosmos, we can cultivate a deeper understanding of our interconnectedness with the natural world and forge a more sustainable and harmonious relationship with the Earth.

As we strive to build a more inclusive and equitable world, let us draw inspiration from the teachings of indigenous cultures and work together to create a future where all beings can thrive in harmony with the cosmos. For in the diversity of human experience lies the potential for greater understanding, compassion, and unity, guiding us on a shared journey of exploration and discovery through the cosmos.

The Spiritual Quest for Cosmic Unity

In the depths of human consciousness lies a profound yearning for connection–to ourselves, to each other, and to the universe that surrounds us. In this chapter, we embark on a spiritual quest for cosmic unity, investigating the role of spirituality and religion in shaping humanity's relationship with the cosmos. We discuss how different religious traditions have sought to understand the mysteries of the universe and explore the ways in which spiritual practices can foster a sense of interconnectedness with the cosmos.

Throughout history, spirituality and religion have played a central role in guiding humanity's search for meaning and purpose in the vastness of the cosmos. From the ancient wisdom of indigenous cultures to the organized religions of the world, human beings have sought to make sense of their place in the universe and to cultivate a deeper understanding of the mysteries that lie beyond the reach of science and reason.

In the rich tapestry of religious traditions, we find a kaleidoscope of beliefs, rituals, and practices that reflect humanity's diverse cultural heritage and spiritual insights. From the monotheistic faiths of Judaism, Christianity, and Islam to the polytheistic traditions of Hinduism and Buddhism, each religious tradition offers its own unique perspective on the nature of existence and the ultimate destiny of the cosmos.

At the heart of many religious teachings lies the concept of cosmic unity—the idea that all beings are interconnected and interdependent, bound together by a shared essence that transcends the boundaries of time and space. In the mystical traditions of Sufism, Kabbalah, and Taoism, seekers of truth have explored the depths of consciousness and experienced moments of profound communion with the divine.

Spiritual practices such as meditation, prayer, and contemplation serve as pathways to this deeper understanding, allowing individuals to transcend the limitations of the ego and experience a sense of oneness with the universe. Through these practices, we cultivate a profound awareness of the interconnectedness of all things and awaken to the inherent divinity that resides within each of us.

Yet, for all its beauty and wisdom, spirituality is not without its challenges. Throughout history, religious differences have been a source of conflict and division, leading to wars, persecution, and oppression. In today's world, we are confronted with urgent moral and ethical questions about the role of religion in shaping our collective destiny and the need to foster greater understanding and tolerance among diverse faith traditions.

As we navigate the complexities of the spiritual quest for cosmic unity, let us embrace the wisdom of the ages and the transformative power of spiritual practice. Let us seek common ground and mutual respect in our shared journey of exploration and discovery through the cosmos. For in the quest for cosmic unity, we discover not only the mysteries of the universe but also the eternal truths that bind us together as members of the human family.

As we continue our exploration of the spiritual quest for cosmic unity, we recognize the profound impact that spirituality has on our individual and collective consciousness. Across cultures and throughout history, spiritual seekers have been drawn to the transcendent mysteries of the cosmos, seeking to connect with something greater than themselves and to find meaning in the vastness of existence.

In the modern world, the quest for cosmic unity takes on new dimensions as we grapple with the challenges of living in an interconnected and rapidly changing global society. In an era marked by technological advancement and scientific discovery, spirituality offers a beacon of hope and a reminder of the sacredness of life in all its forms.

In the face of environmental degradation, social injustice, and existential uncertainty, spiritual traditions offer valuable insights into the interconnectedness of humanity with the Earth and the cosmos. Many religious and spiritual teachings emphasize the importance of stewardship, compassion, and mindfulness as pathways to healing and transformation on both personal and planetary levels.

Through practices such as mindfulness meditation, yoga, and sacred ritual, individuals can cultivate a deeper sense of connection with the universe and align themselves with the rhythms of nature and the cosmos. In moments of quiet contemplation and reflection, we can attune ourselves to the cosmic symphony that surrounds us, listening for the whispers of wisdom that echo through the ages.

Moreover, the spiritual quest for cosmic unity invites us to embrace diversity and celebrate the richness of human experience in all its myriad forms. In a world marked by division and polarization, spirituality reminds us of our shared humanity and the interconnectedness of all life. By honoring the sacredness of diversity and seeking unity in our common humanity, we can build bridges of understanding and compassion that transcend the boundaries of culture, religion, and ideology.

As we journey forward on the spiritual quest for cosmic unity, let us remain open to the mysteries of the universe and the infinite possibilities that lie ahead. Let us cultivate a spirit of reverence and awe for the beauty and complexity of creation, and let us walk in harmony with the rhythms of the cosmos, guided by the light of wisdom and love.

For in the quest for cosmic unity, we discover not only the secrets of the universe but also the timeless truths that unite us as children of the stars, bound together by the sacred threads of existence. And in the journey of the soul, we find not only the answers to our questions but also the eternal journey of exploration and discovery that defines the human experience.

The Poetry of the Stars

In the vast expanse of the cosmos, poetry serves as a beacon of light, illuminating the beauty and wonder that lie beyond the reaches of our imagination. In this chapter, we embark on a journey to explore the universe through the lens of poetry, delving into the timeless verses of cosmic poets and inviting readers to discover their own poetic responses to the majesty of the cosmos.

Poetry has long been a cherished companion on humanity's quest to understand the mysteries of the universe. From the ancient epics of Homer to the celestial sonnets of Shakespeare and the cosmic musings of modern-day poets, the stars and galaxies have inspired countless verses that speak to the human soul.

At the heart of cosmic poetry lies a profound reverence for the beauty and wonder of the universe—a recognition of the interconnectedness of all things and the infinite possibilities that lie beyond the horizon of human understanding. In the shimmering light of distant stars and the swirling mists of nebulae, poets find echoes of their own longing and aspirations, weaving words into tapestries of meaning and metaphor that capture the essence of existence.

Among the pantheon of cosmic poets, few have left a more indelible mark on the literary landscape than the Persian mystic Rumi, whose ecstatic verses dance with the rhythm of the cosmos and the song of the soul. Through his timeless poetry, Rumi invites us to gaze upon the stars with eyes of wonder and to embrace the mystery and magic that surrounds us.

In the footsteps of Rumi, poets throughout history have sought to capture the ineffable beauty of the universe in verse, from the Romantic odes of Wordsworth and Keats to the transcendental hymns of Whitman and Dickinson. Through their words, we are transported to distant galaxies and cosmic realms, where the boundaries of time and space dissolve in the light of poetic imagination.

But the poetry of the stars is not confined to the pages of history; it lives and breathes in the hearts and minds of poets everywhere, waiting to be awakened by the spark of inspiration. In the quiet moments of contemplation and reflection, we are invited to listen to the whispers of the cosmos and to weave our own verses into the fabric of creation.

As we journey through the universe, let us embrace the poetry of the stars and allow its beauty to illuminate our path. Let us open our hearts to the wonder and mystery that surrounds us, and let us celebrate the infinite possibilities that lie within us. For in the poetry of the stars, we discover not only the secrets of the cosmos but also the timeless truths that bind us together as children of the universe, forever reaching for the stars.

In the lyrical cadence of cosmic poetry, we find a reflection of the human spirit reaching out to touch the infinite expanse of the universe. Poets, like celestial navigators, guide us through the depths of space and time, inviting us to ponder the mysteries of existence and to marvel at the grandeur of creation.

One cannot explore the poetry of the stars without encountering the works of Mary Oliver, whose tender verses beckon us to immerse ourselves in the beauty of the natural world. With each line, Oliver invites us to witness the delicate dance of life unfolding across the cosmos, reminding us of our place within the intricate web of existence.

Similarly, the cosmic musings of Carl Sagan echo through the corridors of time, inviting us to contemplate our cosmic insignificance and our profound connection to the universe. In Sagan's words, we find solace in the vastness of space and the timeless beauty of the cosmos, reminding us of the wonders that lie beyond the confines of our earthly existence.

As we journey deeper into the realm of cosmic poetry, we are reminded of the power of language to transcend the limitations of the human experience and to transport us to realms beyond imagination. Through the alchemy of words, poets weave tapestries of meaning and metaphor that capture the essence of the universe and awaken within us a sense of wonder and awe.

Yet, the true beauty of cosmic poetry lies not in the words themselves, but in the emotions and insights they evoke within us. In the quiet moments of contemplation, we are invited to listen to the silent symphony of the stars and to reflect on the eternal mysteries that surround us.

In the end, the poetry of the stars serves as a reminder of our shared humanity and our collective longing for meaning and belonging in the vastness of the cosmos.

Through the timeless verses of cosmic poets, we find solace in the beauty of the universe and inspiration to embark on our own journey of exploration and discovery.

As we gaze upon the stars with eyes of wonder and reverence, let us remember the words of the poets who have come before us, and let us honor their legacy by continuing to seek truth, beauty, and understanding in the boundless expanse of the cosmos. For in the poetry of the stars, we discover not only the secrets of the universe but also the eternal longing of the human soul to be part of something greater than ourselves.

Cosmic Consciousness and Human Evolution

In the unfolding saga of human evolution, the concept of cosmic consciousness stands as a beacon of light, guiding us towards a deeper understanding of our place in the universe. In this chapter, we embark on a journey to examine the role of cosmic consciousness in shaping human evolution and development. We discuss how expanded states of awareness can foster a deeper connection to the universe and explore the potential for humanity to evolve towards a more harmonious relationship with the cosmos.

At the heart of cosmic consciousness lies the recognition that we are not separate from the universe, but rather integral parts of its vast and interconnected web of existence. From the smallest atom to the grandest galaxy, all of creation is imbued with the same divine essence, pulsating with the rhythms of life and consciousness.

Throughout history, human beings have sought to awaken to this higher state of awareness through various spiritual practices, meditation, and contemplation. In moments of expanded consciousness, we transcend the boundaries of the ego and experience a profound sense of unity with all that is, feeling intimately connected to the stars, the planets, and the cosmos itself.

Through the lens of cosmic consciousness, we come to understand that the universe is not just a collection of disparate objects, but a living, breathing organism with its own consciousness and intelligence. Every atom, every molecule, every galaxy is infused with the spark of divinity, expressing itself in an infinite variety of forms and manifestations.

As we awaken to the reality of cosmic consciousness, we begin to see the world with new eyes, recognizing the inherent beauty and interconnectedness of all things. We no longer view ourselves as separate individuals, but as integral parts of a larger cosmic tapestry, each contributing to the symphony of life in our own unique way.

Moreover, expanded states of awareness have the power to transform our relationship with the universe, inspiring us to live in greater harmony with the natural world and with each other. When we see ourselves as interconnected beings, we recognize that the well-being of one is inseparable from the well-being of all, and we are motivated to act with compassion, kindness, and love.

In the grand unfolding of human evolution, the awakening of cosmic consciousness holds the promise of a brighter future for humanity and the planet. As we continue to explore the depths of our own consciousness, we open ourselves to the infinite possibilities that lie within us, and we step boldly into a new era of cosmic evolution.

In the end, the journey towards cosmic consciousness is not just a personal quest, but a collective endeavor that encompasses all of humanity. As we embrace the interconnectedness of all life and awaken to the beauty of the cosmos, we discover that we are not just passengers on this cosmic journey, but active participants in the unfolding drama of creation. And in this realization, we find the seeds of a new world —a world where love, compassion, and harmony reign supreme, and where humanity stands in harmony with the cosmos, united in purpose and vision.

As we delve deeper into the exploration of cosmic consciousness and its impact on human evolution, we encounter profound implications for our collective journey as a species. The awakening of cosmic consciousness offers us a transformative lens through which to view ourselves, our relationships, and our place in the cosmos.

One of the key insights of cosmic consciousness is the realization that the boundaries we perceive between ourselves and the universe are illusory. In truth, we are inseparable from the fabric of existence, woven into the tapestry of the cosmos in ways that transcend our individual identities. This awareness invites us to embrace a sense of interconnectedness and interdependence with all living beings and the natural world.

Furthermore, expanded states of consciousness allow us to tap into deeper layers of wisdom and insight that transcend the limitations of rational thought. In these states, we may experience profound revelations about the nature of reality, the interconnectedness of all things, and our inherent potential for growth and transformation.

Through practices such as meditation, mindfulness, and contemplation, individuals can cultivate a direct experience of cosmic consciousness, opening the doorway to expanded awareness and insight. As we journey inward, we discover hidden reservoirs of wisdom and creativity that enable us to navigate the complexities of life with grace and resilience.

The awakening of cosmic consciousness also holds the promise of catalyzing a paradigm shift in our collective consciousness–a shift towards greater harmony, compassion, and cooperation. As more individuals awaken to the interconnectedness of all life, we are called to reevaluate our relationships with each other and the planet, fostering a deeper sense of stewardship and reverence for the Earth and its ecosystems.

Moreover, the evolution of cosmic consciousness invites us to reconsider our societal structures and systems in light of our interconnectedness with the cosmos. As we recognize the inherent worth and dignity of every individual and the web of life, we are inspired to create communities and institutions that reflect the values of unity, equality, and sustainability.

In the grand tapestry of human evolution, the awakening of cosmic consciousness represents a pivotal moment of transformation and possibility. It calls upon us to embrace our interconnectedness with the cosmos and to harness the power of expanded awareness for the greater good of all beings.

As we continue to explore the depths of cosmic consciousness, let us remember that the journey towards evolution is ongoing—a journey of discovery, growth, and transformation that unfolds with each passing moment. And as we awaken to the infinite potential that lies within us, may we walk boldly into the future, guided by the light of cosmic wisdom and the boundless love that unites us all.

The journey towards cosmic consciousness is not without its challenges and obstacles. It requires us to confront deeply ingrained patterns of thought and behavior that perpetuate separation and division, both within ourselves and in society at large. It calls upon us to transcend the limitations of the ego and to embrace a broader, more inclusive perspective that encompasses the entirety of existence.

In the pursuit of cosmic consciousness, we may encounter resistance and doubt, as well as moments of confusion and uncertainty. Yet, it is precisely in these moments of challenge that the seeds of transformation are sown. It is through the process of facing our fears and transcending our limitations that we come to discover the boundless potential that lies within us.

As we awaken to cosmic consciousness, we begin to recognize the inherent unity that underlies the diversity of existence. We see that we are all expressions of the same divine essence, each playing a unique role in the unfolding drama of creation. This realization fosters a sense of compassion, empathy, and solidarity with all beings, inspiring us to work together towards the common good.

Moreover, the awakening of cosmic consciousness invites us to reevaluate our relationship with the natural world and to adopt more sustainable and harmonious ways of living. As we recognize the interconnectedness of all life, we feel called to honor and protect the Earth and its ecosystems, preserving its beauty and diversity for future generations.

In the grand tapestry of human evolution, the journey towards cosmic consciousness represents a profound leap forward in our collective development. It offers us the opportunity to transcend the limitations of the past and to embrace a new vision of humanity–one rooted in love, wisdom, and interconnectedness.

As we continue on this journey, let us remember that the awakening of cosmic consciousness is not the end goal, but rather a beginning–a beginning of a new chapter in the story of human evolution. It is a journey of exploration, discovery, and growth that unfolds with each passing moment, inviting us to embrace the fullness of our potential as cosmic beings.

In the light of cosmic consciousness, we find the courage to step boldly into the unknown, guided by the wisdom of the ages and the boundless love that animates the universe. And as we walk this path together, may we be filled with hope, inspiration, and a deep sense of reverence for the miracle of existence that surrounds us.

The Gaia Hypothesis and Planetary Consciousness

In the vast expanse of the cosmos, our home planet Earth stands as a jewel of life, a vibrant oasis teeming with diversity and vitality. In this chapter, we delve into the Gaia hypothesis, a revolutionary theory that proposes Earth as a self-regulating, living system, and explore the concept of planetary consciousness. We discuss humanity's role as stewards of the Earth and contemplate the profound implications of the Gaia hypothesis for our understanding of humanity's place in the cosmos.

The Gaia hypothesis, first proposed by scientist James Lovelock and biologist Lynn Margulis in the 1970s, challenges conventional views of the Earth as a passive, inert rock orbiting the sun. Instead, it suggests that Earth is a complex, self-regulating system in which living organisms interact with the atmosphere, oceans, and geology to maintain conditions conducive to life.

At the heart of the Gaia hypothesis lies the idea of planetary consciousness–the notion that Earth is not just a collection of individual organisms, but a single, interconnected organism in its own right. Like the cells of a living body, each species plays a unique role in maintaining the health and balance of the planet, contributing to the overall well-being of the Earth as a whole.

Humanity, with its capacity for reason, creativity, and self-awareness, occupies a special position within the web of life. As stewards of the Earth, we have a unique responsibility to care for and protect the fragile ecosystems that sustain us. Yet, throughout history, our actions have often led to imbalance and destruction, threatening the very systems upon which our survival depends.

In recent years, however, there has been a growing recognition of the need to adopt a more holistic and sustainable approach to our relationship with the Earth. From grassroots movements advocating for environmental conservation to global initiatives aimed at addressing climate change, there is a growing awareness of the interconnectedness of all life and the importance of preserving the health and vitality of the planet.

The Gaia hypothesis offers us a new perspective on humanity's place in the cosmos, reminding us of our interconnectedness with all living beings and the Earth itself. It challenges us to see ourselves not as separate from nature, but as integral parts of a larger, planetary organism—one that is intricately woven into the fabric of the universe.

As we contemplate the implications of the Gaia hypothesis, we are called to embrace a deeper sense of reverence and respect for the Earth and all its inhabitants. We are challenged to reevaluate our priorities and to seek harmony with the natural world, recognizing that our well-being is intimately linked to the health and vitality of the planet.

In the grand tapestry of cosmic evolution, the Gaia hypothesis invites us to see ourselves as caretakers and custodians of a precious and irreplaceable gift—the gift of life on Earth. It reminds us that we are not just passengers on this cosmic journey, but active participants in the unfolding drama of creation, entrusted with the sacred task of nurturing and preserving the beauty and diversity of life for generations to come.

Continuing from the exploration of the Gaia hypothesis and planetary consciousness, we uncover profound insights into humanity's interconnectedness with the Earth and the broader cosmos.

The Gaia hypothesis challenges us to shift our perspective from viewing Earth as a mere collection of resources to recognizing it as a living, breathing entity deserving of our care and respect. It prompts us to reevaluate our relationship with the natural world and to acknowledge the intricate web of life that sustains us.

At its core, the Gaia hypothesis invites us to cultivate a deeper sense of mindfulness and awareness of our impact on the Earth's ecosystems. It calls upon us to consider the consequences of our actions and to strive for a more harmonious and sustainable way of living in accordance with the principles of planetary well-being.

Moreover, the concept of planetary consciousness challenges us to expand our sense of identity beyond national borders, cultural boundaries, and individual interests. It encourages us to embrace a sense of global citizenship and to recognize that we are all interconnected members of a single planetary community.

In a world marked by environmental degradation, social injustice, and political strife, the principles of planetary consciousness offer us a path forward towards greater unity, cooperation, and collective action. They remind us that the challenges we face–from climate change to biodiversity loss–are global in nature and require global solutions rooted in compassion, empathy, and solidarity.

As stewards of the Earth, we have a responsibility to protect and preserve the precious gift of life for future generations. We must work together to address the root causes of environmental degradation and to create a more sustainable and equitable world for all beings.

The Gaia hypothesis serves as a powerful reminder of our interconnectedness with the cosmos and the profound interconnectedness of all life. It calls upon us to awaken to the beauty and wonder of the Earth and to honor the sacredness of all living beings.

In the grand symphony of cosmic evolution, let us embrace the principles of planetary consciousness and strive to live in harmony with the Earth and all its inhabitants. Let us walk gently upon the Earth, guided by the wisdom of the Gaia hypothesis and the boundless love that animates the universe.

For in the unity of all things, we find the true essence of our humanity and the infinite potential for growth, transformation, and renewal. And in the embrace of planetary consciousness, we discover the key to unlocking a brighter future for ourselves, for the Earth, and for generations yet to come.

As we continue to explore the profound implications of the Gaia hypothesis and planetary consciousness, we are reminded of the interconnectedness of all life and the sacred responsibility we bear as stewards of the Earth.

In our quest to align with the principles of planetary well-being, we are called to embrace a holistic approach to sustainability –one that recognizes the interdependence of ecological, social, and economic systems. This requires us to transcend narrow notions of progress and prosperity and to cultivate a deeper appreciation for the intrinsic value of life in all its forms.

At the heart of planetary consciousness lies a deep reverence for the Earth and a recognition of its inherent wisdom and resilience. As we attune ourselves to the rhythms of nature, we come to understand that the Earth is not simply a resource to be exploited, but a living, breathing organism that sustains and nourishes us all.

Through practices such as mindfulness, meditation, and ecological stewardship, we can deepen our connection to the Earth and foster a sense of reverence for the natural world. By slowing down and tuning into the present moment, we can cultivate a greater appreciation for the beauty and wonder that surrounds us, and for the intricate web of life that sustains us.

Moreover, the principles of planetary consciousness challenge us to reimagine our social and economic systems in ways that honor the well-being of both people and the planet. This requires us to move beyond the pursuit of short-term profit and to prioritize the long-term health and vitality of our communities and ecosystems.

In the face of global challenges such as climate change, biodiversity loss, and social inequality, the principles of planetary consciousness offer us a guiding light and a source of hope. They remind us that we are not powerless in the face of adversity, but that we have the capacity to effect positive change through collective action and collaboration.

As we journey forward on the path of planetary consciousness, let us remember that we are all interconnected members of the Earth community, bound together by a common destiny and a shared responsibility to care for our planet and each other. Let us embrace the wisdom of the Gaia hypothesis and strive to live in harmony with the Earth, honoring its sacredness and protecting its precious gifts for generations to come.

In the grand tapestry of cosmic evolution, the principles of planetary consciousness offer us a vision of a world where all beings live in balance and harmony with the Earth, guided by the principles of love, compassion, and respect. And in the embrace of planetary consciousness, we find the keys to unlocking a future of peace, prosperity, and sustainability for ourselves and for all living beings.

Cosmic Perspectives on Ethics and Morality

In the grand tapestry of the cosmos, questions of ethics and morality take on profound significance as we navigate the complexities of human existence. In this chapter, we embark on a journey to explore how cosmic perspectives can inform our understanding of ethics and morality. We delve into the principles of cosmic ethics, including compassion, interconnectedness, and stewardship, and examine how ethical considerations can guide our actions as individuals and as a global society.

At the heart of cosmic perspectives on ethics and morality lies a recognition of the interconnectedness of all life and the inherent dignity and worth of every being in the universe. From the smallest microbe to the grandest galaxy, all of creation is imbued with a spark of divinity, reflecting the infinite diversity and beauty of the cosmos.

Cosmic ethics invites us to embrace a worldview that transcends narrow notions of self-interest and tribalism and recognizes our shared responsibility to care for the Earth and all its inhabitants. It calls upon us to cultivate qualities such as compassion, empathy, and kindness towards all living beings, recognizing that our actions have far-reaching consequences that ripple through the fabric of existence.

Compassion lies at the heart of cosmic ethics, serving as a guiding principle for how we relate to ourselves, to others, and to the natural world. When we approach life with an open heart and a spirit of compassion, we honor the interconnectedness of all life and recognize the inherent worth and dignity of every being.

Interconnectedness is another fundamental principle of cosmic ethics, reminding us that we are all part of a larger web of existence that spans the cosmos. When we recognize our interconnectedness with all living beings, we come to understand that the well-being of one is inseparable from the well-being of all, and we are inspired to act with greater care and consideration for the welfare of others.

Stewardship is the third pillar of cosmic ethics, calling upon us to act as responsible caretakers of the Earth and its ecosystems. As stewards of the planet, we have a sacred duty to protect and preserve the beauty and diversity of life for future generations, ensuring that the Earth remains a thriving and vibrant home for all beings.

As we navigate the complexities of modern life, ethical considerations can serve as a compass to guide our actions and decisions. Whether in our personal relationships, our professional endeavors, or our interactions with the natural world, cosmic ethics offers us a framework for living in alignment with our highest values and aspirations.

Moreover, as a global society facing pressing challenges such as climate change, social injustice, and environmental degradation, ethical considerations take on added significance. They remind us of our shared humanity and our collective responsibility to create a world that is just, sustainable, and compassionate for all beings.

In the grand symphony of cosmic evolution, ethics and morality serve as the guiding principles that shape our journey towards a more enlightened and harmonious future. As we embrace the principles of cosmic ethics–compassion, interconnectedness, and stewardship–we awaken to the inherent beauty and interconnectedness of all life, and we step boldly into a future guided by the light of wisdom and love. As we delve deeper into cosmic perspectives on ethics and morality, we come to realize that our actions and decisions have far-reaching consequences that extend beyond the boundaries of our immediate lives. In the vast tapestry of existence, every choice we make reverberates through the interconnected web of life, shaping the destiny of our planet and all its inhabitants.

Compassion, as a cornerstone of cosmic ethics, invites us to extend our empathy and understanding to all living beings, transcending barriers of species, culture, and ideology. When we approach the world with a heart full of compassion, we recognize the inherent dignity and worth of every individual and strive to alleviate suffering wherever it is found.

Interconnectedness reminds us that we are not isolated islands but integral parts of a larger whole, intricately woven into the fabric of existence. When we honor our interconnectedness with all life, we acknowledge our shared responsibility to care for the Earth and each other, working together towards the common good of all beings.

Stewardship, as a guiding principle of cosmic ethics, calls upon us to act as custodians of the Earth, nurturing and protecting its precious ecosystems for future generations. As stewards of the planet, we recognize that we are entrusted with the well-being of all life, and we strive to make decisions that honor the delicate balance of nature and promote sustainability and harmony.

In our increasingly interconnected world, ethical considerations have become more vital than ever, guiding us in our interactions with one another and with the natural world. As we confront complex challenges such as climate change, social inequality, and environmental degradation, cosmic ethics offers us a roadmap for navigating the complexities of the modern age with wisdom and compassion.

Moreover, as a global society, we are called to embrace the principles of cosmic ethics on a collective level, recognizing that we are all citizens of a shared planet with a common destiny. Through collaboration, cooperation, and mutual respect, we can work together to address the pressing issues facing humanity and create a world that reflects our highest aspirations for peace, justice, and sustainability.

In the grand unfolding of cosmic evolution, ethics and morality serve as guiding stars, illuminating the path towards a more enlightened and compassionate future. As we embrace the principles of cosmic ethics–compassion, interconnectedness, and stewardship–we awaken to the beauty and interconnectedness of all life, and we step boldly into a future guided by the light of wisdom and love.

In the journey of understanding cosmic perspectives on ethics and morality, we find ourselves challenged to not only contemplate but also to embody these principles in our daily lives and interactions. As we continue to explore the depth of these concepts, we uncover new layers of insight and wisdom that inspire us to live with greater integrity, compassion, and purpose.

One of the profound implications of cosmic ethics is the recognition that our individual actions have ripple effects that extend far beyond ourselves. Every choice we make, whether big or small, contributes to the collective fabric of human experience and shapes the course of our shared destiny. Therefore, cultivating awareness and mindfulness in our actions becomes paramount, as we strive to align our behaviors with the highest ethical ideals.

Furthermore, the principles of cosmic ethics invite us to embrace a broader sense of responsibility towards all living beings and the planet as a whole. This expanded awareness calls upon us to consider the well-being of future generations, as well as the intricate interdependence of all life forms. It prompts us to reconsider our relationship with the Earth and to adopt lifestyles and practices that promote harmony, sustainability, and ecological balance.

Moreover, as we navigate the complexities of a rapidly changing world, ethical considerations serve as guiding principles that help us navigate the moral terrain with clarity and integrity. They provide us with a moral compass that directs our choices and empowers us to stand firm in our commitment to justice, compassion, and equality.

In the broader context of global society, the principles of cosmic ethics invite us to engage in meaningful dialogue and collaboration, transcending cultural, political, and ideological divides. By fostering a spirit of understanding and cooperation, we can work together to address the pressing issues facing humanity and to build a more just, equitable, and sustainable world for all.

In essence, the journey of exploring cosmic perspectives on ethics and morality is not merely an intellectual exercise but a call to action—a call to embody the highest ideals of love, compassion, and wisdom in our thoughts, words, and deeds. It is an invitation to awaken to our interconnectedness with all life and to recognize the inherent dignity and worth of every being in the universe.

As we embrace the principles of cosmic ethics and strive to live in alignment with our deepest values, we become catalysts for positive change in the world, contributing to the evolution of human consciousness and the flourishing of life on Earth. In this way, we honor the sacred interconnectedness of all existence and fulfill our role as custodians of the cosmic tapestry of life.

The Dance of the Cosmos: Exploring Cosmic Rhythms

In the vast expanse of the cosmos, a mesmerizing symphony of rhythms and cycles unfolds, weaving together the fabric of existence in an intricate dance of cosmic proportions. In this chapter, we embark on a journey to explore the rhythms and cycles that shape the cosmos, from the graceful orbits of planets to the pulsating rhythms of stars. We delve into how these cosmic rhythms influence life on Earth, from the ebb and flow of the tides to the delicate balance of circadian rhythms. Moreover, we reflect on the interconnectedness of cosmic and earthly rhythms and their profound significance for the human experience.

At the heart of the cosmos lies a profound sense of rhythm—a rhythmic heartbeat that reverberates throughout the universe, guiding the movements of celestial bodies and shaping the course of cosmic evolution. From the graceful dance of planets around their parent stars to the mesmerizing orbits of galaxies in the vast cosmic ocean, these rhythms form the very foundation of existence itself.

On Earth, we witness the echoes of these cosmic rhythms in the natural world around us. The ebb and flow of the tides, the changing seasons, and the rhythmic cycles of day and night—all bear witness to the intricate interplay of cosmic forces that shape our planet and sustain life in all its forms.

Moreover, the rhythms of the cosmos exert a profound influence on the human experience, shaping our perceptions, emotions, and sense of time. From the ancient rhythms of lunar cycles to the pulsating rhythms of the human heart, we are intimately connected to the rhythmic tapestry of the universe, woven into its intricate patterns of light and shadow.

In the study of cosmology and astronomy, scientists seek to unravel the mysteries of these cosmic rhythms, probing the depths of space and time to unlock the secrets of the universe. Through observations, experiments, and mathematical models, they strive to understand the underlying principles that govern the movements of celestial bodies and the dynamics of cosmic evolution.

Yet, beyond the realm of science lies a deeper truth–a truth that transcends the boundaries of intellect and reason. It is a truth that speaks to the very essence of our being, reminding us of our place within the vast cosmic symphony of existence.

As we reflect on the interconnectedness of cosmic and earthly rhythms, we are invited to attune ourselves to the subtle harmonies of the universe, to listen to the whispers of the stars and the rhythms of the Earth. In the quietude of contemplation, we may glimpse the profound beauty and majesty of the cosmos, and find solace in the eternal dance of creation.

In the grand tapestry of cosmic evolution, the rhythms of the cosmos serve as a reminder of the interconnectedness of all life, weaving together the threads of existence in a timeless dance of unity and harmony. And in the embrace of these cosmic rhythms, we may find peace, wonder, and a deeper sense of belonging in the vast and wondrous cosmos. As we delve deeper into the exploration of cosmic rhythms, we uncover profound insights into the interconnectedness of all existence and the exquisite beauty of the universe's dance. At the heart of this cosmic symphony lies a fundamental harmony–a symphony of rhythms that reverberate throughout the cosmos, guiding the movements of galaxies, stars, and planets in a graceful ballet of cosmic proportions.

The orbits of planets around their parent stars are choreographed with exquisite precision, governed by the elegant laws of celestial mechanics. From the majestic sweep of Jupiter to the delicate pirouette of Mercury, each planet follows its own unique path, weaving together a celestial tapestry of breathtaking beauty and complexity.

Moreover, the rhythms of the cosmos exert a profound influence on life here on Earth, shaping the very fabric of our existence in subtle and profound ways. The ebb and flow of the tides, driven by the gravitational pull of the moon and the sun, create a rhythmic dance that shapes coastal landscapes and sustains fragile ecosystems along the shoreline.

Similarly, the changing seasons, driven by the tilt of Earth's axis as it orbits the sun, mark the passage of time and the cycle of life and death in the natural world. From the bursting forth of new life in the springtime to the quiet slumber of winter's embrace, each season carries with it its own unique rhythm, reflecting the eternal cycle of renewal and transformation.

At a deeper level, the rhythms of the cosmos resonate within us, shaping our thoughts, emotions, and perceptions of reality. From the pulsating rhythms of our own heartbeat to the rhythmic patterns of our breath, we are intimately connected to the cosmic symphony that surrounds us, woven into its intricate tapestry of light and sound.

In the contemplation of cosmic rhythms, we are invited to attune ourselves to the subtle harmonies of the universe, to listen deeply to the song of the stars and the whispers of the wind. In the stillness of our hearts, we may hear the echoes of eternity, echoing through the corridors of time and space.

As we embrace the rhythms of the cosmos, we come to recognize our own place within the vast and wondrous dance of creation. We are but fleeting actors on the stage of existence, swept up in the grand sweep of cosmic evolution, yet intimately connected to the very fabric of reality itself.

In the grand tapestry of cosmic evolution, the rhythms of the cosmos serve as a reminder of the interconnectedness of all life, weaving together the threads of existence in a timeless dance of unity and harmony. And in the embrace of these cosmic rhythms, we may find solace, wonder, and a deeper sense of belonging in the vast and wondrous cosmos that surrounds us.

As we continue our exploration of cosmic rhythms, we delve even deeper into the intricacies of the universe's dance, seeking to understand the profound implications of these rhythms for our lives and our understanding of existence itself.

The dance of the cosmos is not merely a spectacle to be observed from afar but a living, breathing expression of the universe's inherent intelligence and creativity. Each celestial movement, each rhythmic pattern, carries with it the imprint of cosmic wisdom, inviting us to contemplate the deeper mysteries of existence and our place within the cosmic tapestry.

In the rhythms of the cosmos, we find echoes of our own inner rhythms–the steady beat of our hearts, the rhythmic flow of our breath, the cyclical patterns of our thoughts and emotions. Just as the planets and stars move in harmonious synchrony, so too are we connected to the larger rhythms of the universe, each of us playing a unique role in the cosmic symphony of life.

Moreover, the rhythms of the cosmos hold profound implications for our understanding of time and space. In the eternal dance of creation, past, present, and future merge into a seamless continuum, transcending the limitations of linear time and inviting us to embrace the timeless essence of existence. In this timeless realm, every moment becomes a portal to eternity, every breath a doorway to infinity.

As we attune ourselves to the rhythms of the cosmos, we come to recognize the sacredness of every moment, the beauty of every movement, the majesty of every celestial dance. In the stillness of our hearts, we find a deep sense of connection to the vast and wondrous universe that surrounds us, a sense of awe and wonder that transcends the boundaries of the known and the familiar.

In the dance of the cosmos, we find a reflection of our own innermost longing–for beauty, for truth, for meaning. And as we immerse ourselves in the rhythms of creation, we discover that we are not separate from the universe but intimately woven into its fabric, part of a larger whole that encompasses all of existence.

In the grand symphony of cosmic evolution, the rhythms of the cosmos remind us of our interconnectedness with all life, of our place within the vast and wondrous dance of creation. And as we embrace these rhythms, we open ourselves to the infinite possibilities of existence, to the boundless creativity and intelligence that animate the universe itself.

Cosmic Healing and Holistic Wellness

In the intricate tapestry of existence, the concept of cosmic healing emerges as a profound pathway to holistic wellness, offering us a deeper understanding of our interconnectedness with the universe and the inherent healing potential that resides within each of us. In this chapter, we embark on a journey to explore the essence of cosmic healing and its transformative power to promote physical, emotional, and spiritual well-being. We delve into practices such as meditation, yoga, and energy healing that cultivate a deeper connection to the universe, and we reflect on the role of cosmic awareness in fostering a state of profound wholeness and balance.

At its core, cosmic healing embraces the fundamental principle that we are not separate from the universe but intricately woven into its vast and wondrous fabric. We are expressions of the cosmic dance of creation, imbued with the same divine essence that animates the stars and galaxies. In recognizing our interconnectedness with the cosmos, we open ourselves to the limitless healing potential that flows through the universe, tapping into a source of wisdom and vitality that transcends the confines of the physical body.

Practices such as meditation serve as powerful gateways to cosmic healing, inviting us to quiet the mind, open the heart, and attune ourselves to the rhythms of the universe. Through the practice of meditation, we cultivate a deep sense of presence and awareness, allowing us to connect with the infinite reservoir of peace and serenity that resides within us. In the stillness of meditation, we become attuned to the subtle energies of the cosmos, aligning ourselves with the flow of universal wisdom and love.

Similarly, yoga offers us a holistic path to cosmic healing, integrating movement, breath, and awareness to harmonize body, mind, and spirit. Through the practice of yoga, we awaken the dormant energies within us, releasing tension and blockages that inhibit the free flow of life force energy. As we move through the graceful sequences of yoga postures, we attune ourselves to the cosmic rhythms that pulse through the universe, finding balance and alignment in every breath and every movement.

Energy healing modalities such as Reiki and acupuncture offer additional pathways to cosmic healing, channeling universal life force energy to restore balance and vitality to the body, mind, and spirit. By working with the subtle energies of the body, these practices help to clear energetic blockages, release stored emotions, and promote healing on all levels of being. As we open ourselves to the flow of cosmic energy, we awaken to a state of profound wholeness and well-being, experiencing a deep sense of connection to the universe and all that is.

In the journey of cosmic healing and holistic wellness, we come to recognize that true healing is not merely the absence of disease but a state of vibrant health and vitality that encompasses every aspect of our being. It is a journey of self-discovery and self-transformation, a journey of awakening to the inherent beauty and perfection that lies within us and all around us.

As we embrace the principles of cosmic healing, we step into a new paradigm of wellness that honors the interconnectedness of body, mind, and spirit. We cultivate practices that nourish and replenish us on all levels of being, fostering a state of radiant health and vitality that emanates from the depths of our being. And in the embrace of cosmic healing, we discover a profound sense of wholeness and harmony that resonates with the very essence of the universe itself.As we journey further into the realm of cosmic healing and holistic wellness, we begin to uncover the profound implications of our interconnectedness with the universe and the transformative power that lies within us. By embracing practices that cultivate a deeper connection to the cosmos, we awaken to the inherent wisdom and healing potential that resides at the core of our being.

In the pursuit of cosmic healing, meditation emerges as a powerful tool for quieting the mind, calming the spirit, and attuning ourselves to the rhythms of the universe. Through the practice of meditation, we create a sacred space for inner exploration and self-discovery, allowing us to access the deep reservoirs of peace, clarity, and insight that reside within us. As we quiet the chatter of the mind and surrender to the present moment, we open ourselves to the infinite possibilities of healing and transformation that lie beyond the realm of thought and perception.

Similarly, yoga offers us a pathway to holistic wellness by integrating movement, breath, and awareness to harmonize body, mind, and spirit. Through the practice of yoga, we cultivate strength, flexibility, and balance in both body and mind, aligning ourselves with the natural rhythms of the universe. As we flow through the graceful sequences of yoga postures and connect with the rhythm of our breath, we awaken to a state of deep relaxation and inner peace, experiencing a profound sense of unity and wholeness that transcends the boundaries of the self.

Energy healing modalities such as Reiki, acupuncture, and sound healing provide additional avenues for accessing the healing power of the universe and restoring balance to the body, mind, and spirit. By working with the subtle energies that flow through the body, these practices help to clear blockages, release stagnant energy, and promote healing on all levels of being. As we open ourselves to the flow of cosmic energy and attune ourselves to the universal life force that surrounds us, we awaken to a state of profound alignment and well-being, experiencing a deep sense of connection to the divine intelligence that animates the universe.

In the journey of cosmic healing and holistic wellness, we come to recognize that true health is not merely the absence of disease but a state of vibrant vitality and well-being that encompasses every aspect of our being. By embracing practices that nourish and replenish us on all levels of being, we awaken to the inherent beauty and perfection that lies within us and all around us. And in the embrace of cosmic healing, we discover a profound sense of wholeness and harmony that resonates with the very essence of the universe itself.

As we continue to explore the mysteries of cosmic healing and holistic wellness, we open ourselves to the infinite possibilities of healing and transformation that lie within us. By embracing the interconnectedness of body, mind, and spirit, we awaken to the inherent wisdom and healing potential that resides at the core of our being, and we step boldly into a future guided by the light of love, compassion, and unity. In the journey of cosmic healing and holistic wellness, we embark on a quest to rediscover our innate connection to the universe and harness its transformative power to nurture and heal our body, mind, and spirit. As we delve deeper into the exploration of cosmic healing practices, we uncover a profound truth–that the universe itself is a vast reservoir of wisdom, energy, and healing potential, waiting to be tapped into by those who are willing to listen and attune themselves to its rhythms.

Meditation, with its timeless practice of stillness and inner reflection, serves as a gateway to the realm of cosmic healing. Through the simple act of sitting in silence and turning our awareness inward, we create a sacred space for communion with the cosmic forces that surround us. In the quiet depths of meditation, we learn to listen to the whispers of the universe, to hear its gentle guidance and feel its nurturing embrace. In this state of deep receptivity, healing becomes not just a physical process but a spiritual journey of self-discovery and awakening.

Yoga, with its graceful movements and mindful breathing, offers us a pathway to harmony and balance in body, mind, and spirit. Through the practice of yoga, we learn to cultivate a deep sense of presence and awareness, as we move through the flowing sequences of postures and connect with the rhythm of our breath. In the union of body and breath, we experience a profound sense of integration and wholeness, as we attune ourselves to the cosmic energies that flow through us and around us. In this state of alignment, healing becomes not just a destination but a way of being–a natural expression of our connection to the universe.

Energy healing modalities such as Reiki, acupuncture, and sound healing offer us powerful tools for accessing the universal life force energy that animates all living things. Through the gentle touch of the hands, the insertion of fine needles, or the resonance of sound vibrations, we activate the body's innate healing mechanisms and restore balance to the subtle energies that flow within us. In the hands of a skilled practitioner, energy healing becomes a sacred ritual–a dance of light and sound, of love and compassion–that opens the door to profound transformation and healing.

As we embrace the principles of cosmic healing and holistic wellness, we come to realize that true healing is not just about curing symptoms or alleviating pain, but about reconnecting with our true nature and aligning ourselves with the cosmic rhythms of the universe. It is about recognizing that we are not separate from the world around us, but an integral part of the cosmic dance of creation–a dance of light and shadow, of birth and death, of love and loss. And in the embrace of this cosmic dance, we find healing, wholeness, and the infinite wisdom of the universe itself.

The Quantum Universe: Exploring Cosmic Mysteries

In the vast expanse of the cosmos lies a realm of mystery and wonder–a realm governed not by the familiar laws of classical physics, but by the enigmatic principles of quantum mechanics. In this chapter, we embark on a journey to explore the intricacies of the quantum universe and unravel its profound mysteries. We delve into the fundamental principles of quantum mechanics and their far-reaching implications for our understanding of the cosmos. We discuss the mysteries of quantum entanglement, superposition, and wave-particle duality, and we reflect on the ways in which quantum physics challenges our notions of reality and expands our cosmic perspective.

At the heart of quantum mechanics lies a profound departure from the deterministic worldview of classical physics–a departure that opens the door to a realm of infinite possibilities and cosmic potential. In the quantum universe, particles exist not as discrete entities with well-defined properties, but as waves of probability, fluctuating in a state of superposition until observed or measured.

One of the most intriguing phenomena of the quantum world is that of entanglement– the mysterious connection that exists between particles, regardless of the distance that separates them. In the quantum realm, particles can become entangled in a state of mutual dependence, such that the measurement of one particle instantaneously affects the state of its entangled partner, even if they are light-years apart. This phenomenon challenges our conventional notions of space and time and hints at a deeper interconnectedness that pervades the fabric of reality.

Another perplexing aspect of the quantum universe is the phenomenon of superposition, wherein particles can exist in multiple states simultaneously until observed or measured. This principle gives rise to the famous thought experiment of Schrödinger's cat—a hypothetical scenario in which a cat enclosed in a box exists in a state of superposition, simultaneously alive and dead, until the box is opened and the cat's fate is determined. Superposition challenges our intuitive understanding of reality and invites us to reconsider the nature of existence itself.

Furthermore, the concept of wave-particle duality lies at the heart of quantum mechanics, suggesting that particles such as electrons and photons can exhibit both wave-like and particle-like behavior depending on the context of observation. This duality blurs the boundaries between the macroscopic and microscopic worlds and underscores the inherent uncertainty that permeates the quantum realm.

As we contemplate the mysteries of the quantum universe, we come to realize that reality is far more complex and nuanced than we could have ever imagined. Quantum physics challenges us to expand our cosmic perspective, inviting us to embrace a world of uncertainty, probability, and infinite potential. It beckons us to peer beyond the veil of classical reality and explore the boundless realms of the quantum cosmos—a realm where the mysteries of existence unfold in a dance of light and shadow, of waves and particles, of entanglement and superposition.

In the exploration of the quantum universe, we embark on a quest for truth and understanding–a quest that leads us to the very edge of human knowledge and beyond. As we peer into the depths of the quantum abyss, we glimpse the infinite majesty and complexity of the cosmos, and we are humbled by the realization that the universe is far more wondrous and mysterious than we could have ever imagined. And in the embrace of this cosmic mystery, we find inspiration, awe, and a profound sense of wonder that transcends the boundaries of space and time.

In our exploration of the quantum universe, we are confronted with the profound realization that reality, as we perceive it, is a tapestry woven from the threads of uncertainty, probability, and interconnectedness. The principles of quantum mechanics challenge our most deeply held beliefs about the nature of existence, inviting us to question the very fabric of reality and to embrace a new paradigm of cosmic understanding.

At the heart of quantum mechanics lies the principle of uncertainty, encapsulated in Heisenberg's famous uncertainty principle. This principle asserts that certain pairs of physical properties, such as position and momentum, cannot be precisely measured simultaneously. Instead, the more accurately we know one property, the less precisely we can know the other. This inherent uncertainty underscores the probabilistic nature of the quantum world, where particles exist in a state of constant flux and unpredictability.

Furthermore, the concept of wave-particle duality challenges our conventional understanding of the nature of matter and energy. In the quantum realm, particles such as electrons and photons exhibit both wave-like and particle-like behavior, blurring the distinction between waves and particles and highlighting the inherent interconnectedness of all phenomena. This duality invites us to reconsider our perception of reality and to embrace the idea that matter and energy are not separate entities but interconnected manifestations of a unified cosmic fabric.

Moreover, the phenomenon of quantum entanglement stands as one of the most baffling and intriguing aspects of the quantum universe. When particles become entangled, their states become inextricably linked, such that the measurement of one particle instantaneously influences the state of its entangled counterpart, regardless of the distance between them. This phenomenon challenges our classical notions of causality and suggests the existence of a hidden web of interconnectedness that transcends the boundaries of space and time.

As we grapple with the mysteries of the quantum universe, we are drawn into a realm of infinite possibility and cosmic potential. The quantum world is a realm where uncertainty reigns supreme, where particles can exist in multiple states simultaneously, and where the very act of observation shapes the fabric of reality itself. In this quantum dance of creation, we glimpse the profound beauty and complexity of the cosmos, and we are humbled by the realization that the universe is far more intricate and wondrous than we could have ever imagined.

In the exploration of the quantum universe, we are invited to embrace a new way of seeing–a way of seeing that transcends the limitations of classical physics and opens the door to a deeper understanding of the interconnectedness of all things. As we peer into the quantum abyss, we are reminded of the boundless mysteries that lie at the heart of existence, and we are inspired to continue our journey of cosmic exploration, ever seeking to unravel the secrets of the universe and unlock the infinite potential that lies within us all. As we delve deeper into the quantum universe, we encounter the tantalizing prospect of a reality that defies our conventional understanding–a reality where particles can exist in multiple states simultaneously, where the very act of observation alters the fabric of existence, and where the boundaries between the macroscopic and microscopic worlds blur into obscurity. In this realm of quantum mysteries, we are confronted with the profound implications of quantum mechanics for our understanding of the cosmos and our place within it.

The principles of quantum mechanics challenge us to expand our cosmic perspective and to embrace a new way of thinking about the nature of reality. They invite us to question the assumptions and limitations of classical physics and to embark on a journey of exploration into the depths of the quantum abyss.

One of the most remarkable features of the quantum universe is the phenomenon of superposition, wherein particles can exist in a multitude of states simultaneously until observed or measured. This principle challenges our intuition and invites us to contemplate the existence of parallel realities–alternate universes where every possible outcome of a quantum event is realized. In the realm of superposition, the universe becomes a vast tapestry of possibilities, where every choice, every decision branches off into an infinite number of potential futures.

Another intriguing aspect of the quantum universe is the phenomenon of quantum entanglement, wherein particles become inextricably linked in a state of mutual dependence, regardless of the distance that separates them. This phenomenon suggests the existence of a hidden web of interconnectedness that permeates the fabric of reality, transcending the limitations of space and time. In the dance of entangled particles, we glimpse the profound unity and interconnectedness of all things, and we are reminded of the interconnected nature of the cosmos itself.

Moreover, the principles of quantum mechanics challenge our conventional notions of causality and determinism, suggesting that the universe may be far more fluid and unpredictable than we could have ever imagined. In the quantum universe, the very act of observation becomes an integral part of the process of reality creation, shaping the outcomes of quantum events in ways that defy our classical understanding of cause and effect.

As we contemplate the mysteries of the quantum universe, we are reminded of the boundless potential and infinite creativity that lie at the heart of existence. In the quantum dance of creation, we glimpse the wondrous beauty and complexity of the cosmos, and we are inspired to continue our quest for understanding, ever seeking to unravel the secrets of the universe and unlock the mysteries that lie beyond the veil of perception.

Cosmic Citizenship: Embracing Our Role in the Universe

In the grand tapestry of the cosmos, humanity stands as a sentinel species, endowed with the remarkable capacity for self-awareness and conscious reflection. As inhabitants of the universe, we are not merely passive observers but active participants in the unfolding drama of cosmic evolution. In this chapter, we delve into the concept of cosmic citizenship and explore humanity's responsibilities as custodians of the cosmos. We examine how cultivating cosmic awareness can inspire environmental stewardship, social justice, and global cooperation, and we reflect on the importance of embracing our role as cosmic citizens in the face of planetary challenges and opportunities.

At its core, cosmic citizenship is about recognizing our interconnectedness with the universe and embracing the inherent responsibilities that come with being inhabitants of the cosmos. It is about acknowledging that our actions have consequences not only for ourselves and our fellow human beings but for the entire web of life that sustains us. It is about understanding that we are stewards of the Earth and trustees of its future, charged with the sacred task of preserving and protecting the precious gift of life for generations to come.

One of the key aspects of cosmic citizenship is environmental stewardship–the recognition that we are custodians of the Earth and guardians of its natural resources. In an age of environmental crisis, it is imperative that we take proactive measures to safeguard the planet and mitigate the impact of human activities on the biosphere. By adopting sustainable practices, conserving biodiversity, and combating climate change, we honor our commitment to the Earth and affirm our role as responsible custodians of its fragile ecosystems.

Moreover, cosmic citizenship extends beyond the realm of environmental stewardship to encompass broader issues of social justice and global cooperation. In an increasingly interconnected world, we are called upon to recognize the inherent dignity and worth of every human being and to work towards a more equitable and inclusive society. By promoting human rights, fostering dialogue, and addressing systemic inequalities, we strive to create a world where all individuals can flourish and thrive, regardless of their background or circumstances.

Furthermore, cultivating a sense of cosmic citizenship requires us to transcend narrow tribal loyalties and embrace a planetary perspective that transcends borders and boundaries. In an age of globalization, we are confronted with the urgent need to forge a sense of global solidarity and cooperation in the face of shared challenges such as pandemics, poverty, and conflict. By recognizing our common humanity and working together towards common goals, we can harness the collective wisdom and resources of humanity to address the pressing issues of our time and build a more just, sustainable, and peaceful world for future generations.

In conclusion, cosmic citizenship is not merely a theoretical concept but a practical ethic that calls upon us to live in harmony with the universe and to fulfill our responsibilities as stewards of the Earth. It is a call to action—a call to embrace our role as cosmic citizens and to work towards a brighter, more hopeful future for ourselves and for all living beings. By embracing the principles of cosmic citizenship, we can harness the transformative power of cosmic awareness to create a world that reflects the inherent beauty, diversity, and interconnectedness of the cosmos itself.

As we delve deeper into the concept of cosmic citizenship, we recognize that it entails not only a recognition of our interconnectedness with the universe but also a profound sense of responsibility for the well-being of all life on Earth. It is a call to transcend narrow self-interest and embrace a broader vision of planetary stewardship that reflects our shared destiny as inhabitants of a common home.

At the heart of cosmic citizenship lies the understanding that we are not separate from the cosmos but integral parts of its intricate web of existence. From the depths of the oceans to the heights of the mountains, from the smallest microorganisms to the most complex ecosystems, every living being plays a vital role in the cosmic symphony of life. By acknowledging our interconnectedness with all living beings, we cultivate a deep sense of reverence and respect for the diversity and beauty of the natural world.

Environmental stewardship is a central tenet of cosmic citizenship, encompassing a wide range of practices and initiatives aimed at preserving and protecting the Earth's fragile ecosystems. From reducing our carbon footprint to advocating for renewable energy solutions, from conserving biodiversity to restoring degraded habitats, there are countless ways in which we can contribute to the health and vitality of the planet. By taking concrete actions to safeguard the environment, we honor our commitment to future generations and ensure that they inherit a world that is rich in biodiversity and teeming with life.

In addition to environmental stewardship, cosmic citizenship also calls upon us to address broader issues of social justice and equity. In a world marked by inequality and injustice, we are called upon to stand in solidarity with the marginalized and oppressed, to amplify their voices, and to work towards a more just and equitable society for all. By challenging systems of oppression and advocating for the rights of the most vulnerable among us, we embody the principles of cosmic citizenship and strive to create a world where every individual can live with dignity, freedom, and equality.

Furthermore, cosmic citizenship requires us to embrace a global perspective that transcends national boundaries and fosters cooperation and collaboration among nations. In an era of globalization, the challenges we face–from climate change to global pandemics–are inherently global in nature and require coordinated responses at the international level. By fostering dialogue, building bridges of understanding, and promoting mutual respect and cooperation, we can harness the collective wisdom and resources of humanity to address the pressing challenges of our time and build a more peaceful and sustainable world for future generations.

In conclusion, cosmic citizenship is a call to action–a call to embrace our interconnectedness with the universe and to fulfill our responsibilities as stewards of the Earth. It is a recognition that we are all citizens of a common planet, bound together by the fragile threads of life. By embracing the principles of cosmic citizenship and acting with compassion, wisdom, and foresight, we can create a world that reflects the inherent beauty, diversity, and interconnectedness of the cosmos itself.

As we continue to explore the profound concept of cosmic citizenship, we delve deeper into the multifaceted responsibilities and opportunities it presents to humanity. Through the lens of cosmic awareness, we recognize that our individual actions ripple outward, shaping the fabric of the universe and influencing the course of cosmic evolution. With this awareness comes a profound sense of duty–a duty to act as custodians of the cosmos and stewards of the Earth, nurturing and preserving the delicate balance of life for future generations.

Environmental stewardship stands as a cornerstone of cosmic citizenship, reminding us of our interconnectedness with the natural world and our responsibility to protect and preserve its wonders. By embracing sustainable practices in our daily lives, conserving resources, and advocating for policies that prioritize environmental conservation, we demonstrate our commitment to the well-being of the planet and all its inhabitants. Through initiatives such as reforestation, wildlife conservation, and sustainable agriculture, we work to restore balance to ecosystems and safeguard the biodiversity that sustains life on Earth.

Social justice and equity are also integral components of cosmic citizenship, calling upon us to address systemic inequalities and injustices that undermine the dignity and rights of individuals around the world. By advocating for equality, inclusivity, and human rights, we strive to create a more just and compassionate society where every person has the opportunity to thrive and fulfill their potential. Through acts of solidarity, empathy, and allyship, we stand in solidarity with marginalized communities and work to dismantle barriers that perpetuate discrimination and oppression.

Furthermore, cosmic citizenship fosters a spirit of global cooperation and collaboration, recognizing that the challenges we face–from climate change to global health crises–are inherently interconnected and require collective action at the international level. By fostering dialogue, building bridges of understanding, and promoting diplomacy and peacebuilding, we strive to create a world where nations can work together in harmony to address shared challenges and promote the common good. Through initiatives such as international aid and development, humanitarian assistance, and peacekeeping efforts, we extend a hand of friendship and solidarity to our fellow global citizens, fostering a spirit of mutual respect and understanding that transcends borders and boundaries.

In the face of planetary challenges and opportunities, cosmic citizenship reminds us of our shared humanity and our shared destiny as inhabitants of a common home. It calls upon us to embrace our role as custodians of the cosmos and to act with wisdom, compassion, and foresight in shaping the future of our world. Through the principles of cosmic citizenship, we harness the transformative power of cosmic awareness to create a world that reflects the inherent beauty, diversity, and interconnectedness of the universe itself.

The Symphony of the Cosmos: Exploring Cosmic Harmony

In the vast expanse of the cosmos, there exists a profound and intricate harmony–a symphony of cosmic forces that reverberates throughout the universe, binding all things together in a tapestry of interconnectedness and beauty. In this chapter, we embark on a journey to explore the concept of cosmic harmony and to unravel the mysteries of order and chaos that shape the fabric of reality. We delve into the timeless wisdom of ancient and modern philosophies, seeking to find meaning and purpose in the cosmic dance of existence.

At the heart of cosmic harmony lies the recognition of the interconnectedness of all things—a recognition that everything in the universe is intimately linked and interdependent. From the smallest atom to the largest galaxy, from the tiniest microorganism to the most complex ecosystem, every element of the cosmos plays a vital role in the unfolding drama of creation. Just as the notes of a symphony come together to form a harmonious melody, so too do the myriad elements of the universe converge to create a symphony of cosmic proportions.

Yet, within this grand symphony, there exists a delicate balance between order and chaos–a dance of opposing forces that gives rise to the dynamic complexity of the cosmos. Order provides the framework within which life can flourish, offering stability, structure, and predictability to the universe. From the elegant orbits of planets to the intricate patterns of snowflakes, order is woven into the very fabric of existence, guiding the rhythms and cycles of cosmic evolution.

At the same time, chaos serves as a creative force–a catalyst for innovation, adaptation, and transformation. In the swirling maelstrom of chaos, new possibilities emerge, old patterns dissolve, and the boundaries of the known universe are pushed ever outward. Chaos reminds us of the inherent unpredictability and mystery of existence, inviting us to embrace the unknown and to venture boldly into the uncharted territories of the cosmos.

Throughout history, humanity has sought to find meaning and purpose in the cosmos, drawing inspiration from the timeless wisdom of ancient and modern philosophies. From the mystical teachings of the East to the rational inquiries of the West, philosophers, scientists, and sages have explored the nature of reality and the mysteries of existence, seeking to uncover the underlying harmony that unites all things.

In the ancient wisdom traditions of the world, we find echoes of cosmic harmony–in the sacred texts of the Vedas, the teachings of Taoism, and the philosophy of Stoicism. These traditions remind us of our interconnectedness with the universe and offer guidance on how to live in harmony with the rhythms of nature and the cycles of existence.

In the modern scientific era, we continue to explore the mysteries of cosmic harmony through the lens of physics, astronomy, and cosmology. From the elegant equations of quantum mechanics to the awe-inspiring beauty of distant galaxies, science reveals the intricate patterns and symmetries that pervade the cosmos, inviting us to contemplate the profound unity and diversity of the universe.

As we contemplate the symphony of the cosmos, we are reminded of our place within the vast tapestry of existence and our role as stewards of the Earth. Through embracing the principles of cosmic harmony, we cultivate a deep sense of reverence and respect for the interconnected web of life that sustains us, and we strive to live in harmony with the rhythms of the universe. In the embrace of cosmic harmony, we find meaning, purpose, and a profound sense of belonging in the grand symphony of creation.

The exploration of cosmic harmony invites us to delve deeper into the intricate interplay between order and chaos, recognizing that within the apparent duality lies a deeper unity–a unity that transcends the limitations of human understanding. As we contemplate the symphony of the cosmos, we come to understand that the forces of order and chaos are not opposing forces but complementary aspects of a greater whole, working in concert to shape the unfolding tapestry of existence.

Order provides the stability and structure upon which the cosmos is built, offering a framework within which life can flourish and evolve. From the laws of physics that govern the motion of celestial bodies to the intricate patterns of DNA that encode the blueprint of life, order permeates every aspect of the universe, guiding the rhythms and cycles of cosmic evolution. It is the foundation upon which the symphony of the cosmos is built–a timeless melody that resonates through the fabric of space and time.

Yet, within the harmonious structure of order lies the creative chaos of the universe –a boundless sea of potentiality from which all things emerge and into which all things eventually return. Chaos is the driving force of creation, the crucible in which new worlds are forged and old paradigms are shattered. It is the source of innovation, adaptation, and transformation, inviting us to embrace the unknown and to venture boldly into the uncharted territories of the cosmos.

In the midst of the cosmic symphony, humanity has long sought to find meaning and purpose in the rhythms of existence, drawing inspiration from the wisdom traditions of the world and the insights of science and philosophy. From the ancient civilizations of Mesopotamia and Egypt to the modern explorations of quantum mechanics and cosmology, humanity has endeavored to unravel the mysteries of the universe and to uncover the underlying harmony that unites all things.

In the mystical traditions of the East, we find echoes of cosmic harmony–in the teachings of Buddhism, the practices of meditation, and the contemplative arts of yoga and tai chi. These traditions offer insights into the nature of reality and the interconnectedness of all things, inviting us to cultivate inner harmony and balance in our lives.

In the rational inquiries of the West, we find a different approach to understanding cosmic harmony–in the scientific method, the pursuit of knowledge, and the quest for understanding. Science reveals the underlying patterns and symmetries that pervade the cosmos, offering a glimpse into the grandeur and complexity of the universe.

As we contemplate the symphony of the cosmos, we are called to embrace our role as custodians of the Earth and stewards of the cosmic dance. Through cultivating a deep sense of reverence and respect for the interconnected web of life that sustains us, we honor the harmony of the universe and strive to live in harmony with the rhythms of creation.

In the embrace of cosmic harmony, we find meaning, purpose, and a profound sense of belonging in the grand symphony of existence–a symphony that echoes through the cosmos, uniting all things in a timeless dance of light and shadow, of order and chaos, of harmony and beauty. As we continue our exploration of cosmic harmony, we are drawn deeper into the mysterious interplay between order and chaos, discovering that within the seeming chaos of the universe lies a hidden order—a cosmic symphony that transcends our understanding.

Order, with its structured patterns and predictable rhythms, forms the foundation upon which the universe is built. From the graceful dance of planets around stars to the intricate geometry of snowflakes, order permeates every level of existence, guiding the movements of celestial bodies and shaping the tapestry of reality. It is the organizing principle that brings coherence to the cosmos, providing a framework within which life can thrive and evolve.

Yet, amidst the order, there exists a dynamic interplay of chaos—a primordial force of creativity and renewal that gives rise to the endless diversity and complexity of the universe. Chaos is the wellspring of innovation and adaptation, the driving force behind the emergence of new forms and structures. It is the fertile ground from which new ideas, new galaxies, and new civilizations emerge, inviting us to embrace the uncertainty and unpredictability of existence.

In the pursuit of cosmic harmony, humanity has sought wisdom and insight from diverse philosophical and spiritual traditions. From the ancient wisdom of indigenous cultures to the modern insights of quantum physics, each tradition offers a unique perspective on the interconnectedness of all things and the underlying unity of the cosmos.

In the mystical traditions of the East, we encounter the concept of oneness–the idea that all of existence is interconnected and interdependent. From the teachings of Advaita Vedanta to the practices of Zen Buddhism, these traditions emphasize the essential unity of all life and invite us to transcend the illusion of separateness that divides us from one another and from the universe.

In the Western philosophical tradition, we find echoes of cosmic harmony in the writings of Plato, Aristotle, and the Stoics. These philosophers recognized the inherent order and purposefulness of the cosmos, viewing it as a reflection of divine intelligence and wisdom. They saw in the harmony of the universe a source of moral and ethical guidance, guiding humanity towards a life of virtue and excellence.

In the modern scientific era, our understanding of cosmic harmony has been expanded and deepened by the insights of quantum physics, relativity, and cosmology. From the elegant equations of Einstein's theory of relativity to the strange and wondrous world of quantum mechanics, science reveals the hidden symmetries and patterns that underlie the fabric of reality, inviting us to marvel at the beauty and complexity of the universe.

In conclusion, the exploration of cosmic harmony is a journey of discovery–a journey that leads us to the very heart of existence and invites us to contemplate the mysteries of the cosmos. As we immerse ourselves in the symphony of the universe, we come to realize that we are not separate observers but integral participants in the cosmic dance of creation. In the embrace of cosmic harmony, we find solace, inspiration, and a deep sense of belonging in the grand tapestry of existence.

The Cosmic Dance of Creation and Destruction

In the grand theater of the cosmos, creation and destruction are not isolated events but integral aspects of a dynamic and ever-evolving cosmic dance. In this final chapter, we embark on a profound exploration of the interplay between creation and destruction in the universe, delving into the cosmic forces that shape the evolution of stars, galaxies, and planetary systems. We reflect on the parallels between these cosmic processes and the cycles of change in human life and society, inviting readers to contemplate the deeper meanings of existence and the mysteries of cosmic transformation.

At the heart of the cosmic dance lies the eternal interplay of creation and destruction –a delicate balance between the forces of genesis and dissolution that give rise to the breathtaking diversity and complexity of the universe. From the fiery birth of stars in the depths of interstellar clouds to the cataclysmic collisions of galaxies in the depths of space, the cosmos is a theater of ceaseless change and transformation, where old worlds are destroyed to make way for new ones.

The process of stellar birth and death serves as a powerful metaphor for the cosmic dance of creation and destruction. In the crucible of stellar nurseries, vast clouds of gas and dust coalesce under the force of gravity, giving rise to the birth of new stars and planetary systems. Yet, even as stars are born, they are destined to one day meet their end in spectacular displays of cosmic fireworks–collapsing under the weight of their own gravity to form black holes, or exploding in violent supernova explosions that seed the cosmos with the building blocks of life.

Similarly, the evolution of galaxies–the cosmic islands of stars and gas that populate the universe–is shaped by the interplay of creation and destruction. Over billions of years, galaxies collide and merge, giving rise to new forms and structures in the cosmic landscape. In the fiery crucible of galactic collisions, stars are born and worlds are destroyed, leading to the formation of new galaxies and the birth of new cosmic wonders.

The parallels between cosmic processes and the cycles of change in human life and society are striking. Just as stars are born and die, civilizations rise and fall, leaving behind echoes of their existence in the sands of time. The forces of creation and destruction shape the destiny of nations and empires, giving rise to periods of renewal and rebirth amidst the ashes of destruction.

In the face of cosmic uncertainty, humanity is called upon to embrace the inevitability of change and to find meaning and purpose in the midst of chaos. By recognizing the transient nature of existence and embracing the wisdom of impermanence, we can navigate the ebb and flow of cosmic forces with grace and resilience, finding strength in the knowledge that the dance of creation and destruction is an essential aspect of the cosmic order.

As we contemplate the cosmic dance of creation and destruction, we are reminded of our place within the grand tapestry of existence and our role as stewards of the cosmic drama. Through embracing the transformative power of change and honoring the cycles of creation and destruction, we can participate more fully in the cosmic dance, contributing our own unique melody to the symphony of the universe.

In the embrace of the cosmic dance, we find solace, inspiration, and a profound sense of connection to the vast and wondrous cosmos that surrounds us. As we journey onward, may we carry with us the wisdom of the ages and the courage to embrace the mysteries of existence, knowing that we are all part of the eternal dance of creation and destruction.

As the cosmic dance continues, it beckons us to explore the intricate interplay of forces that govern the universe, inviting us to contemplate the profound mysteries of existence and the timeless rhythms of creation and destruction.

In the cosmic realm, creation and destruction are not merely opposing forces but essential aspects of a dynamic and interconnected cosmic web. Each moment of creation gives rise to the possibility of something new, while each moment of destruction paves the way for transformation and renewal. It is a dance of cosmic proportions, where the energies of creation and destruction intermingle and weave together to shape the fabric of reality.

One of the most awe-inspiring manifestations of the cosmic dance is the process of stellar evolution. Stars are born in vast clouds of gas and dust, where the forces of gravity pull matter together to form dense cores. As these cores collapse under their own weight, they ignite into brilliant balls of light and heat, illuminating the cosmos with their radiant energy. Yet, even as they blaze brightly, stars are constantly engaged in a battle against gravity, which seeks to pull them inward and crush them under their own weight.

For some stars, this battle ends in a spectacular explosion known as a supernova, where the outer layers of the star are blown away in a fiery display of cosmic fireworks. In the aftermath of a supernova, the remaining core may collapse to form a neutron star or, in the case of very massive stars, a black hole—a cosmic abyss from which no light can escape.

Yet, even in the midst of destruction, the seeds of creation are sown. The debris ejected by supernovae contains the heavy elements necessary for the formation of new stars, planets, and ultimately, life itself. In this way, the death of one star becomes the birth of countless others, perpetuating the cycle of creation and destruction on cosmic scales.

The dance of creation and destruction is not confined to the realm of stars and galaxies but extends to every corner of the cosmos, shaping the evolution of planets, moons, and even the very fabric of space and time itself. From the violent eruptions of volcanoes to the gradual erosion of mountains by wind and water, from the cataclysmic collisions of asteroids to the gentle embrace of tectonic plates, the forces of creation and destruction sculpt the landscapes of worlds and shape the course of cosmic history.

In the human realm, the dance of creation and destruction is reflected in the rhythms of life and the cycles of change that govern our existence. From the birth of civilizations to the fall of empires, from the rise of new ideas to the decline of old beliefs, humanity is constantly engaged in a process of renewal and transformation—a process that mirrors the eternal dance of creation and destruction unfolding in the cosmos.

As we contemplate the cosmic dance of creation and destruction, we are reminded of our place within the vast and wondrous tapestry of existence. We are but fleeting beings in the grand scheme of the cosmos, yet our actions have the power to shape the course of cosmic history. In embracing the mysteries of creation and destruction, we open ourselves to the wonders of the universe and the infinite possibilities that lie beyond.

As we gaze upon the stars and contemplate the mysteries of the cosmos, may we be filled with a sense of wonder and awe at the beauty and complexity of the universe. May we find solace in the eternal rhythms of creation and destruction, knowing that we are all part of the cosmic dance–a dance that transcends time and space, and connects us to the very heart of existence itself.

As we continue our journey through the cosmic dance of creation and destruction, we are drawn deeper into the mysteries of existence, exploring the timeless rhythms that shape the fabric of reality.

In the cosmic symphony, creation and destruction are not isolated events but rather interconnected aspects of a larger cosmic process. They are the yin and yang of the universe, the light and shadow that dance together in an eternal embrace. From the birth of stars to the collapse of galaxies, from the formation of planets to the extinction of species, the forces of creation and destruction are ever-present, shaping the evolution of the cosmos in ways both profound and beautiful.

One of the most striking manifestations of the cosmic dance is the phenomenon of cosmic evolution—the gradual unfolding of the universe from its primordial beginnings to the present day. Over billions of years, galaxies have formed, stars have ignited, and planets have coalesced from the dust of interstellar space. Yet, even as new worlds emerge, old ones are constantly being reshaped and transformed by the relentless forces of cosmic change.

In the depths of space, galaxies collide and merge, giving rise to stunning displays of cosmic fireworks that illuminate the night sky. These collisions are not acts of destruction but rather acts of creation, as they trigger the formation of new stars and fuel the growth of supermassive black holes at the centers of galaxies. Through these cataclysmic events, the universe continually reinvents itself, giving birth to new forms and structures that defy our wildest imagination.

On a smaller scale, the cosmic dance is reflected in the evolution of planetary systems, where worlds are born and die in a constant cycle of creation and destruction. From the molten surfaces of newly formed planets to the barren landscapes of ancient worlds, each celestial body bears the scars of its cosmic history, telling a story of upheaval and transformation that spans billions of years.

In the human realm, the cosmic dance finds expression in the cycles of life and death that shape our existence. From the moment of our birth to the moment of our passing, we are immersed in a dynamic process of growth, change, and renewal–a process that mirrors the eternal rhythms of the universe itself. Just as stars are born and die, civilizations rise and fall, leaving behind a legacy that echoes through the ages.

Yet, amidst the chaos and uncertainty of cosmic evolution, there is a profound sense of order and purpose that permeates the universe. It is the guiding force that directs the cosmic dance, orchestrating the movements of galaxies, stars, and planets with exquisite precision. It is the unseen hand that shapes the destiny of the cosmos, weaving together the threads of creation and destruction into a tapestry of unimaginable beauty and complexity.

In the end, the cosmic dance of creation and destruction is not merely a spectacle to behold but a profound mystery to be embraced–a mystery that invites us to contemplate the deeper truths of existence and the interconnectedness of all things. As we gaze upon the wonders of the universe, may we be filled with a sense of wonder and awe at the majesty of creation, knowing that we are all part of the cosmic dance–a dance that transcends time and space, and connects us to the very heart of existence itself.

Cosmic Connections" is a captivating journey through the boundless wonders of the universe, each chapter delving deeper into the mysteries that shape our understanding of existence. From the vast expanse of space to the depths of the human soul, we explore the profound connections that link us to the cosmic tapestry unfolding around us.

Through the lens of scientific inquiry, we unravel the intricate patterns and forces that govern the cosmos, from the birth of stars to the evolution of galaxies. With each discovery, we gain insight into our place in the cosmic order and marvel at the beauty and complexity of the universe.

In parallel, philosophical reflections invite us to contemplate the deeper truths that underlie our existence. We ponder questions of purpose, meaning, and the nature of reality, seeking to unravel the mysteries that have captivated human minds for centuries.

Artistic expression offers another avenue of exploration, as we discover how the wonders of the cosmos have inspired poets, painters, musicians, and writers throughout history. Through their works, we glimpse the awe and wonder that the universe inspires, transcending language and culture to touch the depths of the human spirit.

Finally, spiritual exploration invites us to delve into the realms of the unseen, where faith and belief intersect with the mysteries of the universe. Drawing upon ancient wisdom traditions and modern insights, we seek to uncover the sacred truths that illuminate our path and guide us on our journey through life.

In the grand cosmic dance of existence, we are but fleeting participants, yet our quest for understanding and connection is timeless. "Cosmic Connections" invites us to join hands and embark on a journey of discovery–a journey that transcends space and time, uniting us all in the vast and wondrous tapestry of the cosmos.

About the Author

Laura Lee's journey from a small town upbringing to becoming a multi-talented author and web designer is a story of inspiration and unwavering determination.

Born and raised in a quaint, close-knit community, Laura was drawn to the mysteries of life and the vast world hidden behind the pages of books from a young age. Her insatiable curiosity and thirst for knowledge were evident even in her early years.

As she matured, Laura's educational pursuits took her to several universities, where she delved into various subjects that captured her imagination. Yet, it was in the year 1999 that her life took a transformative turn. It was during this time that she discovered the captivating realm of computers and the boundless potential they held.

The realization that computers could be a gateway to unleashing her creativity and innovative spirit led Laura to make a pivotal decision. In pursuit of her newfound passion, she enrolled in an esteemed art institution to specialize in web design.

This decision marked the beginning of an incredible journey filled with both challenges and achievements.

During her studies in web design, Laura's creative prowess flourished. She combined her love for visual aesthetics and her growing technical expertise to craft captivating and user-friendly websites. Her designs began to gain recognition and accolades, establishing her as a talented web designer in her own right.

However, Laura's story didn't end with her success in the field of web design. Alongside her flourishing career, she nurtured her love for words and storytelling. Her innate gift for weaving narratives into enchanting tales led her to become an author. Her books, filled with rich storytelling and imaginative worlds, found a devoted readership.

Laura Lee's journey is a testament to the power of pursuing one's passions and the remarkable things that can be achieved through a dedicated and inquisitive spirit. She serves as an inspiring example of how one can evolve from a small-town dreamer into a multifaceted creative professional.

www.ingramcontent.com/pod-product-compliance
Lightning Source LLC
Chambersburg PA
CBHW021450150726
47989CB00001B/472